FINDING THE WAY THROUGH
WATER
On becoming waterwise with the Bible

ROLAND K PRICE

authorHOUSE®

AuthorHouse™ UK Ltd.
1663 Liberty Drive
Bloomington, IN 47403 USA
www.authorhouse.co.uk
Phone: 0800.197.4150

Published by AuthorHouse   04/26/2014

ISBN: 978-1-4918-9518-4 (sc)
ISBN: 978-1-4918-9534-4 (hc)
ISBN: 978-1-4918-9519-1 (e)

Library of Congress Control Number: 2014903405

To Thea

Contents

Preface

Water has fascinated scientists and farmers, cooks and bottle washers, children and grown-ups down the ages. Every part of society has reason to celebrate the existence and uses of water, whether in manufacturing, energy production, agriculture, food preparation, personal hygiene or recreation. Most religions regard water as being vitally important in religious practice, and in helping growth in personal faith. This is particularly true of Christianity and the Bible. We find when reading the Bible that water has many strong associations with God, especially in the ways he has dealt with humankind. This book sets out to explore those associations, and hopefully to enable those unfamiliar with the Bible to catch a glimpse of the ways in which God has made himself known to us, and to encourage them to explore a personal relationship with God our Father and his Son Jesus Christ.

But before we can begin to explore the implications of water for us and our communities, it would be helpful to explore its properties[1]. Water is a strong contender to qualify as the weirdest of substances. We must have water to survive, but 75% of the water on our planet is undrinkable. Water has all the appearances of being mundane: we drink it, wash in it, and cook with it. For the first nine months of our existence, each of us was immersed in water in our mother's womb. But after birth, if we breathe too much water into our lungs, we drown. Water is critical to both life and death.

The chemical name of water is dihydrogen oxide because water molecules are formed when two atoms of explosive hydrogen combine with an atom of highly reactive oxygen: strangely, this is the substance with which we put out fires. Water exists in one or other of its three major states: ice (solid), water (liquid) or vapour (gas), for a wide range of temperatures (say, -80°C to +55°C) at the Earth's surface. If we tried to predict the properties of water from what we know about combinations of hydrogen with other chemical elements near oxygen in the periodic table, we would conclude that water has a boiling point of -73°C (rather than 100°C). So, in some ways water is a mystery. Of its two component elements, hydrogen is the lightest and most abundant chemical element in the universe, contributing about 75% of its mass. Hydrogen is, however, rarely found on the Earth in its elemental state; instead, it is tied up with oxygen as water and with other elements such as carbon to produce hydrocarbons such as oil. Hydrogen is well named, for 'hydro' comes from the Greek for 'water' and 'gen' for 'generator'; it is the 'water generator'. So what is it that makes water so special?

A water molecule has an electrical charge difference across it. This gives the molecule a polar structure (like a tiny magnet), which results in water having the highest *surface tension* of all non-metallic liquids. Surface tension ensures that there is a definite boundary between water and the air. Because of this, a river or lake surface supports skimming stones; Barnes Wallis, of 'dam-buster' fame, was able to bounce his bombs on the water surface in reservoirs above certain German dams during the Second World War[2]. Some insects, such as water boatmen, can 'walk' on the water surface, as can certain small animals if they run fast enough. (We too could 'walk' on water if we were able to run at 64 miles or more than 100 km per hour!). Wind drags the water surface, and through the surface tension generates waves with some fascinating characteristics, such as breaking, whether as 'white caps' on the open ocean or as plunging breakers on a sloping beach, and travelling huge distances across the oceans, carrying information about their generation by the wind, but barely moving the water below them. Surface tension holds each raindrop separate and yet enables a group of drops to combine as they flow down the window pane.

The polar structure of the molecule makes water *adhesive*; that is, it wets and sticks to surfaces. The combination of surface tension and adhesion enables the capillary rise of water in narrow tubes against the force of gravity, which is the way that plants draw up water from the soil to all their cells.

Another unique property of water is that each molecule has up to four so-called hydrogen bonds which help to link the molecules together. The bonds mean that water has unusually high *melting and boiling points* compared with hydrogen sulphide, say; considerably more energy is needed to break the bonds between the water molecules to change their state from ice to water or water to vapour. What is more, apart from ammonia, water has the highest *specific heat capacity* (that is, the amount of heat needed to raise the temperature of water by one degree Celcius) of any substance. This is very important for the Earth's climate, because the water in the oceans moderates the temperature in the atmosphere, making it tolerable for life over most of the planet. The boiling point of water depends on the (air) pressure surrounding it. We are familiar with water boiling at 100°C at sea level, but it will boil at only 68°C at the top of Mount Everest, and amazingly, water is still a liquid, not a vapour, at several hundred degrees centigrade in the geothermal vents at the bottom of deep oceans where the pressure is intense.

Yet another peculiar feature of water is that it is almost impossible to *compress* it. Suppose we fill a plastic bag with water at the ocean surface, seal the bag and take it 4 km down in a deep trench. At such a depth, the pressure is enormous, yet there is at most a 2% decrease in volume of the plastic bag and its contents. The incompressibility of water becomes an important basic assumption when developing the mathematical equations that describe the flow of water in rivers, oceans or urban water systems at the Earth's surface. Yet the (small) compressibility of water can lead to pressure waves in the water pipes of a house when a tap is turned on, producing a knocking sound known as the 'water hammer' effect.

Another consequence of the hydrogen bonds is that water has a maximum *density* at about 4°C. This unique property of water is vital for life in bodies of water; when cooled by air at the surface, water freezes at 0°C; meanwhile the warmer, denser water below at 4°C, sinks. The ice covering the surface takes up a volume about 9% greater than the equivalent amount of water. This ensures that the ice floats and aquatic life can continue to exist under it—so our goldfish don't die! Icebergs, such as one that the Titanic hit during her fateful voyage, have about 11% of their volume showing above the surface. Because ice has a larger volume than the water it replaces, the generation of ice in rock crevices can split the rock. This, together with the erosive property of water flowing in rivers and streams and over the ground surface, means that water and ice have been significant in shaping the topography of the land over geological time.

Due to the salts contained in sea water, water at the surface of the ocean actually freezes at about -1.9°C rather than 0°C. Ice on the sea surface is relatively free of salts because the crystalline structure of water and brine are different. Where ice forms at the north and south poles, the water beneath the ice has a marginally higher concentration of salts, and this generates currents deep down that move the saltier water away from the poles.

We have seen that ice forms on the earth's surface when the air temperature is cold enough; similarly, water vapour in the atmosphere condenses on dust nuclei as air temperature drops. If the temperature of the air is sufficiently cool, then water in the atmosphere becomes a wonderful collection of crystals, which we call snowflakes, each of which is unique. Each snowflake has a different structure, with its own beauty. Eskimos are able to recognize and have words for more than thirty different kinds of snow[3]. We still do not fully appreciate the amazing properties of the water molecule that result in such variety.

The presence of salts in the ocean shows that water is an excellent *solvent*; indeed, it is sometimes referred to as the universal solvent. When it rains, salts are leached out of the soil by the rainwater, and the salts dissolved in the water are eventually transported to the ocean. Water is also especially good at dissolving carbon dioxide from the atmosphere; it is therefore a major player in the fight to reduce the effects of carbon emissions on global warming. Raindrops in the atmosphere and water in the oceans absorb carbon dioxide to form dilute carbonic acid; in other words, normal rain is slightly acidic, as are the oceans. But as we have become all too well aware from oil spills in Alaska and the Gulf of Mexico, water is not that good at dissolving oils and fats, although it transports them very effectively in suspension or on the water surface.

Water vapour in the atmosphere defines *humidity*. If you take a shower or bath and the temperature of the air remains about the same, the vapour can reach the pressure at which it experiences a phase change and condenses out as steam. This is what happens to water in the atmosphere which condenses as the temperature drops, creating fog or dew.

Water is a fluid. When a force is applied to it, although there may be a resistance to the force, the water is deformed and does not return to its original shape. In practice, most fluids have some resistance to being deformed. A measure of this resistance is called 'viscosity'. Treacle and glycerin are very viscous fluids as they show a much stronger resistance to being deformed than water. Viscosity determines the influence of a boundary on the flow of water, while ensuring that there is effectively no flow actually at a solid boundary, such as the bed of a river or the inside of a cup. (You can test this simply by stirring a cup of tea and then noticing what happens to the tea near the side of the cup). In addition, we observe that water in the natural environment rarely flows smoothly; instead it has large swirls and smaller swirls that appear to be random. This is the phenomenon of *turbulence*, which gives fast flowing rivers their frightening appearance.

In contrast to the turbulent flow of rivers, water in a smooth pond or lake may appear to be perfectly still. But this is deceptive, because at the micro scale the water molecules are continually and rapidly moving against each other. This can be observed by very gently putting a drop of dye into a basin of still water. After a short time the dye spreads out due to the *Brownian movement*, or the molecular motion of the water. Of course, turbulence, if it is generated, for example by stirring, is far more effective in dispersing the dye.

There are indeed some strange physical and chemical properties of water. In this book we explore some of the properties and functions of water on earth, and their consequences for life on our planet. Two burgeoning problems that face humankind are the present phase of global warming in our ever changing climate, and the continuing exponential growth in the world's human population when there is no corresponding increase in the amount of fresh water available: there is no more fresh water on the planet than there was 4000 years ago. The crunch is inevitably coming, and although each country has to find its own solutions to the problems of global warming, population growth and water shortage—with developing countries facing the most acute situations—permanent solutions that invoke social justice can only be found if countries work together[4]. Water is a unique substance whose physical and chemical properties appear to have been designed by God for specific tasks in supporting life, and it has a unique and practical part to play in the outworking of God's plan's for the future of the human race through the good news of his Son Jesus Christ.

This book consists of forty short chapters exploring the relationship between that most amazing fluid substance we call 'water' and the Bible. The idea for the book arose from my desire to explore the part that water plays in the biblical accounts of creation and redemption. It offers the possibility of bringing aspects of modern day science and technology to bear on what we experience and know about creation, and similarly of opening up a wealth of insight into the nature of our redemption in Christ through the many related Biblical references to water. This book is meant to provoke discussion as well as inform; it is not intended to be a devotional guide, though it does contain support for devotion.

A note about myself: I studied mathematics at University, but I have spent most of my professional life working in the water sector of civil engineering. I was ordained as a non-stipendiary Deacon and then Priest in the Church of England in 1995-6, but I also continued my professional career and was appointed Professor of Hydroinformatics at UNESCO-IHE in Delft, The Netherlands from 1997. My background therefore, is primarily in science and engineering: I am not a professional theologian. Whereas I am aware that there is an enormous wealth of scholarship, studies and insights into the Bible, (which is itself a library of books written by many different authors and contributors in a number of languages and styles, and with wide ranges of purposes and contexts), for the purpose of this book I deliberately regard the text of the Bible as being complete and sufficient to present God's story. This I accept as a Christian believer: for me the Bible is the 'Word' of God. However, I recognise that many people have questions about the reliability of the text and its application, and I acknowledge that there is both a need to research the nature of the original text(s) and to interpret those texts for today. Where I think it is important in the book to raise questions about the text or its interpretation I do so, but with such a small book it is impossible to include all the critical textual comments and theological queries that one may wish to consider.

The book is designed to be read a chapter a day for six weeks, such as during the Christian period of Lent. You do not need an expert knowledge of water management or a detailed understanding of the Bible. The chapters introduce you to some significant themes of the Bible identified with water. I hope that what is written in this book will encourage you to read the biblical text, and to explore more of your own responsibility for the world in which we live. I trust that my professional colleagues in water engineering and management will find the book relevant in helping them to explore how the Christian faith may relate to their work.

The title for the book is taken from a small book on John's gospel: 'Finding the Way Through John' by John Fenton, whose lectures on the gospels as a whole I was privileged to attend when I was training for ordination. The title of John Fenton's book has more than one implicit meaning involving a play on the words 'Way' (the first Christians were said to follow the Way) and 'John' (the author of the book and the Gospel of Apostle John). I have adapted the title to 'Finding the Way through Water'. It is my hope that readers will find the Way through their encounter with this amazing substance we call 'water'. The term 'waterwise' was coined by me at a time when my Dutch colleagues at UNESCO-IHE had insisted on using the alternative phrase 'wise water' in a publicity document. Since then others have independently picked up the same phrase, which is used in the sub-title of this book.

Roland K Price
Den Haag Autumn 2013

Acknowledgements

I was very fortunate to be introduced to the art of Chris Los-Baxter. Having seen some of her work, I asked if she would be willing to do some etchings for this book. She kindly agreed. I am very grateful for her contribution.

I have been privileged to work with many wonderful people as colleagues in the UK Water Industry and at HR Wallingford between 1972 and 1997. Thereafter I was an academic at UNESCO-IHE, where I was fortunate to work with research colleagues in The Netherlands and with a number of PhD students from around the world. Research has always been for me an interdependent process whereby new ideas and results come from building on what others have discovered. Therefore I am grateful to all those with whom I have worked. To name everybody concerned would take many pages; instead allow me to mention a few key people: they include Paul Samuels, Richard Kellagher and many other colleagues from HR Wallingford and Wallingford Software (as was); Prof. Khurshid Ahmad, Prof. Mike Abbott, Prof. Jean Cunge and Prof. Philip O'Kane; Jan Luijendijk, Prof. Dimitri Solomatine, and other colleagues from UNESCO-IHE; my fourteen PhD students from Delft who gave me some of the most exciting and fulfilling times of my professional life; and Arnold Lobbrecht, Slavco Velickov and other colleagues at HydroLogic. My Christian walk has been influenced by many also, including Bob and Ted Carlos, Rev Graham Pulkingham, Rev Philip and Nan Pare, Canon Vincent Strudwick, and the many personal friends in the congregations of St Mary's, Cholsey, Diocese of Oxford, and of St John and St Philip's, The Hague, Diocese in Europe.

I am especially grateful to Bishop Geoffrey Rowell, Rev. Michael Sanders, Professor David Butler and Mrs Anastasia Hacopian, who read the first draft of this book, commented on it and made suggestions for improvement. The final draft was helpfully reviewed by Ms Trudy Selter and Rev. Ruan Crew. Any poor theology, misquotations, or incorrect interpretations are my responsibility alone. Where the bible is quoted I have consistently used the New International Version, except where indicated separately.

Finally, this book has emerged in the context of my family, my wife Thea, our sons and daughters, their spouses and children. I am deeply indebted to all of them for their patience when holidays were interrupted by something to do with water, for the times I was away from home on business, for spending more time on my computer than was good for them—or me. I am especially grateful to Thea for her support and encouragement, for the many hours she has spent in patiently editing and correcting the text, and for helping me to modify my sometimes terse and awkward scientific style of writing.

1

Water crisis?

'Water, water, everywhere, nor any drop to drink . . . .'
'The Rime of the Ancient Mariner'
by Samuel Taylor Coleridge

Read Job 26:5-14

What crisis? Too much or too little water? In the Netherlands, people know all about water, and generally there is too much of it. They live with the fact that about 26% of their land area is below sea level, and this is where 21% of the population lives. But with ample rainfall over the whole country, and sizeable flows in the Rhine and Maas rivers, Dutch society's need for fresh water is assured, while government initiatives strengthen sea and tidal river defences against any short to medium term rise in sea levels due to global warming. There are many other countries that are not so fortunate. Rising sea levels threaten to wipe the Maldives off the map, and to reduce dramatically the useable land area of Bangladesh, which is also threatened by the increasing frequency of catastrophic floods in the rivers from India flowing through the country. We are all too aware of the devastation caused by hurricane surges in the Gulf of Mexico, by super-storms along the eastern seaboard of North America and through tsunamis generated by earthquakes under the Indian and Pacific oceans.

So is the problem too much water? For the Ancient Mariner, there was both too much and too little. The irony is that about 71% of the earth's surface is covered with water, but it is of little direct value to us because we cannot drink sea water due to the large amount of salts dissolved in it. If we did drink it our bodies would quickly seize up: we need fresh water. We can extract the salts from seawater using desalination, but that process requires large amounts of energy. The sun does this far more efficiently on a global scale than we can hope to do by our puny efforts using our present best technology.

Water from the oceans is evaporated by the sun's radiation to become vapour in the atmosphere. Because of this and the particular properties of water, the oceans are critical in regulating the earth's climate. The water vapour they contribute to the atmosphere is by far the most significant of the greenhouse gases, which act to trap energy from the sun's radiation thereby increasing the global average temperature at the earth's surface. Even small changes to this temperature can make a considerable difference to the climate. Other greenhouse gases include carbon dioxide and methane. Since the Industrial Revolution, human activity has led to a progressive increase in carbon dioxide in the atmosphere. Governments and media have been burdening us with the possible consequences of this increase for global warming, which is threatening to raise sea levels several metres during the next hundred years with devastating consequences for island states such as the Maldives, or delta areas around the world. Fortunately, the oceans and vegetation absorb significant amounts of carbon dioxide from the atmosphere.

Yet there is a far more serious issue than global warming of which we are only dimly aware. This is the burgeoning growth in the human population of our planet. Each of the currently six billion people has basic physical needs in order to survive. In particular, we need air to breathe, water to drink and food to eat. Without air we can survive for at most several minutes. So far as water is concerned, each of us needs a *minimum* of 50 litres of freshwater daily to sustain our bodies and life style[5]. Our bodies are made up of between 50% to 70% water. Without fresh water, our vital organs will not work properly and dehydration sets in; we can survive at most for several days.

A great proportion of our daily food (without which we can live for at most several weeks) comes from agriculture and industrial processing. But agriculture and industry are themselves huge consumers of freshwater. In Africa and Asia, 85% of freshwater withdrawn from natural sources is used for agriculture and 5% by industry[6]. In Europe however, water is used differently: 48% of water is used by industry, while agriculture accounts for 35%. Europe's domestic use of water is 16% compared with about 10% in Africa and Asia. In addition, we all need to remember that we are not the only inhabitants of the planet: we share the earth with many other species which are equally dependent on fresh water for their survival. What is more, plants and animals on land have a continuous need of fresh water to survive and flourish. And flourish it must if we are to sustain our environment and help to ensure the long term future of the human race.

In our developed world, we take the supply of drinkable water from our taps for granted. In particular, Europeans and North Americans are much more profligate in their use of freshwater. We turn on the tap to wash ourselves and prepare our food. We use dishwashers and washing machines to wash our dishes and clothes. We clean our cars, water our gardens, flush our toilets and wash our windows and floors. And the water we use evaporates as if by magic or disappears down the plug hole or the toilet. Beyond the house, the wastewater is 'out of sight and out of mind', because in most developed countries, there are sophisticated systems for treating waste water. But in rural areas of developing countries, piped wastewater collection systems are rare and the waste water is casually discharged into the surrounding environment and ignored. We may be aware of this practice; what is perhaps not such common knowledge is that piped water distribution systems are also rare in many rural areas of developing countries. Approximately two thirds of the world's population get their freshwater from public standpipes, community wells, rivers and lakes, or from rainfall collected from roofs. In rural areas, women and girls have to walk many kilometres, spending hours each day fetching water for their households. To make matters worse, the world is becoming predominantly more urbanised (for the first time in 2008, more than 50% of the world's population was living in cities[7]) putting enormous pressure on inadequate water supply systems, while the demand for food makes agriculture depend more and more on irrigation. Water is being withdrawn from some rivers, lakes and groundwater faster than these sources can be renewed or charged. It is estimated that by 2025, 1.8 billion people will be living in regions of absolute water scarcity while two-thirds of the world's population will be living under water stress conditions[8]. Meanwhile, we are actually reducing our freshwater sources by polluting them with our waste: untreated domestic sewage, toxic industrial effluents and harmful chemicals used by agriculture; and most of these are barely degradable in the environment.

The Millennium Development Goals formulated by UN member states in 2001 aim to spur development in the world's poorer countries during the decade 2005-2015[9]. Target 7C aims to halve the 0.8 billion people without sustainable access to safe drinking water and the 2.3 billion people in urban areas without basic sanitation by 2015. As stated above, a minimum amount of 50 litres of water per person per day is recommended to meet the four basic human needs of drinking, sanitation, bathing and cooking[10]. It is a matter of social justice that we endeavour to meet the targets in the MDGs.

As Christians, we have been entrusted by God with the care of our planet, including its water. In Genesis, the accounts of creation make it very clear that human beings are authorized to be stewards of creation. We are to be fruitful and increase in number. We are to fill the earth and subdue it. We are to 'rule' over the fish of the sea, the birds of the air, and over every living creature that moves over the face of the ground (Genesis 1:28). But this does not mean that we can exploit the resources of the earth while ignoring their sustainability. Although we can manage water in our artificial urban networks intended for supply and drainage, water in the environment is largely outside human management, as the writer of Job highlights in the response of God to Job's speeches to his 'comforters'.

Disobedience to God has prejudiced the relationship between humankind and the productivity of the land. Human arrogance and conflict have abused the synergy between society and the natural environment. We have indeed overexploited the earth's resources. It has been estimated that if everybody in the world were to consume resources at the rate we do in Europe and North America, we would need three earths to satisfy everybody. Obviously, the rate of consumption of the earth's resources by the developed world is unsustainable. Although we have the potential to manage our constructed urban environments, they are vulnerable to miscommunication and societal disintegration. The water crisis is therefore one of immediate concern to all countries, especially in regard to cities. We need to act now before it is too late.

Or has the deadline passed already . . . . ?

We will explore the implications and the consequences of the water crisis in this book. In the meantime, what do you think we can we do as people of faith to address the issues?

Use a search engine to find further information on the web

Resource questions:

Water needs to be equitably distributed to all life, but the natural distribution of water is highly irregular. What should we do as individuals and churches to ensure that all life has a fair share of clean water and that wastewater is properly disposed of?

Use Google or another search engine to find Internet sites concerned with water. What do you think are the most important issues about water for our world today?

Look up the websites of some significant non-government organizations working in the water sector, such as WaterAid, AquaForAll, ARocha and International Water Association (IWA). Identify their main areas of interest and their biggest challenges. How can you become better briefed about the problems and opportunities?

2

Water on Earth

'All the water that will ever be is, right now.'
                    National Geographic, October 1993

Read Psalms 65:9-10

Almost 70% of our planet is covered by water and ice. This is a much bigger proportion of water than habitable land; perhaps we should call our planet 'water' rather than 'earth'! Astrophysicists think that a significant amount of the water on our planet originated from the large number of asteroids and other bodies that crashed into the earth in the past. Almost all of this water is concentrated at or near the surface of the planet, and especially in the oceans. But much of the water is salty—fresh water accounts for only about 2.72% of the total water on the planet. Of this fresh water, 1.76% is in the form of ice and snow, 0.76% is groundwater (in or under the ground), 0.02% is water on the land surface (lakes and rivers) and 0.001% is found in vapour, clouds and precipitation[11]. Most of the ice exists in the polar ice cap above Antarctica, with a much smaller amount on Greenland. It has been estimated that if all the ice at the poles, on Greenland and in the glaciers, were to melt, the oceans would rise by several tens of metres[12]. This possibility would be so serious for many countries, such as The Netherlands, that it warrants a careful study of the effect of any global warming of the climate on sea levels.

The movement of water on the earth's surface is complex. The motion is driven by several forces: the gravitational force of the earth, the gravitational pull of the sun and the moon, the Coriolis force generated by the rotation of the earth, differences in the density of the water due to variations in temperature (affected largely by thermal radiation from the sun) and salinity, and the exchange of energy between water and air pressure and wind fluctuations in the atmosphere. Other important forces resist the motion of the water. The most important of these is the shear stress generated in the flow of the water by random swirls associated with the turbulence. This shear stress can be imagined (very imperfectly) in terms of small slabs of water sliding against each other and resisting their motion. The strength of the shear stress is in part dependent on the viscosity. In this way energy stored in the movement and flow of the water is lost. It is important to remember that water in motion always has the shear stress acting within it. The stress is likely to be greatest at or near a fixed boundary such as a river channel bed or a coastline. These boundaries not only deflect the flow of water coming towards them, but also ensure that the velocity of the flow at each fixed boundary is zero. As an example of the importance of the shear stress internal to the flow of water, consider a river channel with water flowing along it. Because the bed of the channel slopes downhill, the water in the channel is 'pulled' down the channel by the force of gravity. If there was no force opposing the pull of gravity downhill, the water would continue accelerating, getting faster and faster. But we observe that a river normally flows steadily downhill, which means that energy in the flow is being lost. This occurs because the 'rough' texture of the channel boundary, which is made up of, say, gravel, sand or vegetation, induces turbulence in the flow with the result that energy is lost due to the resulting shear stress.

The large ocean currents, which have such an influence on our weather, are caused by the interaction between the Coriolis force generated by the rotation of the earth on its axis, the location of the continents, and the differential input of heat from the sun between the equator and the polar latitudes. Smaller, more localised currents are formed by the periodic fluctuations in the gravitational pull of the moon and the sun. These fluctuations produce the familiar rising and falling of the tides at the seaside, usually twice a day. We might even be aware that in general there is a monthly cycle of tides that coincides with the different phases of the moon.

Even these monthly cycles vary in the longer term due to the variation in the combined effects of the sun with the moon. At an even smaller scale, the waves breaking on our beaches generate local currents, such as the littoral drift along a beach. Water can store considerable amounts of heat before it changes state into water vapour. This ability to absorb heat enables warm ocean currents, such as the Gulf Stream and the Kuroshio Current, flow north from the equator along the east coast of North America and Asia respectively, giving up their heat slowly to the atmosphere in the more northern latitudes. This provides Europe and North West Canada with their temperate climates. Similarly, there are return currents flowing back towards the equator underneath these warm currents, formed by the down-welling of slightly more dense, colder water produced at the poles. Another well-known ocean current is the El Niño current, which is an abnormal warming of the ocean surface in the Eastern Pacific that flows towards the South American coast. The strength of El Niño in terms of its temperature has been found to be a good indicator of the forth-coming weather around the world[13]. There is, of course, a close and complex interaction between the heat in the oceans and the meteorology of the atmosphere; this does not mean that the strength of El Niño causes the weather.

The oceans are, in fact, a huge reservoir of heat. They also provide water vapour (without the salts) to the atmosphere. This water vapour is carried into the upper layers of the atmosphere where the temperature is lower, and the vapour condenses on dust particles to form clouds at different levels. There is an important phenomenon in the upper atmosphere (which is experienced by international airplane travelers), known as the jet stream. This band of fast flowing air has a strong influence on regions of high and low pressure, cloud movement, and especially the 'hot' and 'cold' weather fronts. Meteorologists think that the position and wind speed of the jet stream above the UK were important factors contributing to the record breaking rainfall and flooding during 2012. Under particular conditions, the water drops in the clouds coalesce and freeze to form ice particles. These particles can fall as hail, or snow crystals, or melt into raindrops, which precipitate towards the earth's surface. Most precipitation falls on the oceans but much still falls on the land and at the poles. There is a large range of precipitation intensity at the earth's surface due to the variable cloud in the atmosphere and to the different heat exchange processes going on in the various cloud formations.

Much of the rainfall experienced in Western Europe is generated by frontal systems rotating anti-clockwise around regions of low pressure. Monsoons in the sub-tropics have a similar structure. But a convection cell in a thunderstorm can be a violent source

Water on planet earth

of rainfall, especially when moist water laden air is forced to rise up mountain slopes, cooling as it goes. The most violent storms recorded in recent history in India have been responsible for almost 1 metre depth of rainfall in 24 hours. This depth approaches the maximum amount of water that scientists think can be stored in the atmosphere above a given location. Forecasting rainfall is very important for society, but is fraught with uncertainty. The reliability of a forecast deteriorates rapidly into the future because the motion of the air, including water vapour, carbon dioxide and other gases in the atmosphere, is a 'chaotic' process; that is, the evolution of the atmosphere, as characterized by variables such as temperature, pressure, density and velocity (wind), is extremely sensitive to the initial conditions. A very small change to the initial conditions can make a huge change to the resulting rainfall. Rainfall consists of fresh water, which may be contaminated by acids formed by contact with other chemicals in the atmosphere. The freshness of the water is vital for animal and plant life on the land[14].

In the Bible, rainfall is seen to be significant. Particularly striking is the comparison drawn between Egypt, with its very low annual rainfall, and Israel, which is watered directly by the 'early and the latter' rains. Egypt has been dependent historically on water from the Nile, which comes from thousands of kilometres to the south. The Ancient Egyptians raised water manually from the river into irrigation canals; to do this, farmers used their feet to tread the lifts for raising the water (see Deuteronomy 11.10). In contrast, Israel, with its moderate rainfall is the 'land that the LORD loves' with its annual early and latter rains watering the hills and valleys. The whole population, whether just or unjust, benefits from the rainfall. God is good, and sends rain to benefit all.

The evaporation of water vapour from the oceans is but one part of the hydrological 'cycle'. As rain falls, it wets any surface, and it may fall on trees, vegetation, roads and buildings. As the rainfall increases, water may infiltrate through the ground cover into the soil underneath. Plant roots take up some of this water, raising it through capillary action and osmosis to all cells in the plants. Most of this water transpires through the plant leaves and is evaporated. Meanwhile the remaining infiltrated water proceeds through the voids in the soil and may reach an underground aquifer. Of course, if there is too much rainfall on the ground for it all to be intercepted or to infiltrate, excess water may form small ponds on the surface and then flow over the ground surface following the line of greatest slope.

The topography of the ground surface determines where the first significant streams appear. These streams are fed by the surface runoff and water from the soil, possibly including the groundwater aquifer. Smaller streams flow into bigger streams, until the resulting flow can be sufficient for the channel to be termed a 'river'. Ultimately, the water in this network of streams and river channels flows to a lake or to the sea. The flow in a river is determined largely by gravity, and the resistance to the flow by the boundary of the river channel. What is more, over thousands of years, the flows erode the soil and sediment to form channels that have pronounced banks, separating the channels from any associated flood plains.

When there is hardly any rainfall on the catchment draining to a river, a residual flow is normally retained in the channel due to slow drainage of rainfall water from the soil and the groundwater aquifer in the rock strata below the surface of the catchment. This residual flow is termed the 'base' flow by hydrologists. Given enough intense rainfall however, the water runs off quickly over the surface, providing a much more rapid contribution to the flow in the streams. Where there is no rainfall for a prolonged period, the stream may dry up completely, as in desert wadis. These are ephemeral streams that are only active following a brief but possibly heavy rain storm occurring once a year or even less frequently. Where the flow generated in a river is so significant that it overflows its banks; the river is said to be in 'flood'. As the rainfall intensity generating the flood is random, there is a risk for people living or working on flood plains, a risk that many communities are prepared to take in view of the preferential access to fresh water, opportunities to farm the rich alluvial soils, and the advantages of good lines of communication along the valleys.

On the other hand, there are many places, such as in Egypt, where the rainfall is negligible; in the absence of any other water source these places are deserts where life is generally dormant, waiting for the rare rainfall that can make the desert bloom. Often rainfall in such places is torrential when it occurs, bringing ephemeral streams briefly to life, recharging underground aquifers and replenishing wells and oases. The world of the Bible is largely one of limited rainfall, which implies that human and animal populations tend to be focused near good sources of fresh water, such as rivers, wells and oases, and rich agricultural soil where food is relatively easy to grow. As a consequence, the availability or lack of water, play an important part in the outcomes of the history of Israel.

Resource questions:

What major incidents can you think of in the Old Testament that are significantly dependent on water?

What has water done in fashioning the landscape you are familiar with?

What recent natural disasters have you been aware of which were caused by or involved water?

3

Water and Creation I

'In the beginning God created the heavens and the earth.'

Genesis 1:1

Read Genesis 1:1-23

The opening words of the Bible in Genesis, quoted above, are filled with drama. In Psalm 104 we find a fuller poetic description of creation and the place of water in it. Scholars think this psalm was composed before Genesis was written down, and the understanding of what we call nature as described in the psalm mirrors that of Mesopotamian and Canaanite thought some five millennia ago. The psalm focuses on the conferring of order on what was originally a state of chaos. God is seen as being in full control of what he has created: he 'makes the clouds his chariot' (v.3b); he 'covered (the earth) with the deep as with a garment; the waters stood above the mountains' (v.6); he assigned a place for the waters (v.8); he 'set a boundary they cannot cross' (v.9); he 'makes springs pour water into the ravines' (v.10); they provide water to animals and a secure environment for birds (v.11, 12). God 'waters the mountains from his upper chambers' so that the earth is satisfied (v.13) and can be fruitful to the benefit of all living things, especially gladdening the heart of man. The Psalmist is lost in wonder at the nature of the world around him, and can everywhere see the hand of God in all that is. In Psalm 8:4 the Psalmist asks a basic question for each of us: 'What is man that you think of him, mere man that you care for him?' In other words: 'Why are we here?' This is a poignant question when men and women are referred to in another psalm as 'a puff of wind' or 'a passing shadow'; see Psalm 144:4. But the implication of Psalm 8 is that man is 'Godlike' in his nature, and especially in relation to the created order.

In his second letter, the Apostle Peter states that the world has changed little since creation, and it is likely to continue that way. He points out that 'long ago by God's word the heavens existed and the earth was formed *out of water and by water*'; see 2 Peter 3.5. It can also be destroyed by water, as with the story of Noah. Another Old Testament passage, Proverbs 8:22-31, talks about Wisdom being present at the beginning of creation 'When there were no oceans . . . no springs abounding with water . . . when he (God) established the clouds above and fixed securely the fountains of the deep, when he gave the sea its boundary so that the waters would not overstep his command . . . Then I (Wisdom) was the craftsman at his side. I was filled with delight day after day, rejoicing always in his presence, rejoicing in his whole world and delighting in mankind.'

The act of creation isn't restricted to the beginning of the world. Throughout the Bible, God is the Creator with 'Wisdom' at his side (Proverbs 8:22-31) which some commentators identify with Jesus Christ. The Hebrew verb, *bārā*, 'to create' comes six times in Genesis 1, but, centuries later, in his preaching to the exiles in Babylon, Isaiah speaks of God creating things seventeen times; see, for example, Isaiah 40:26, 41:20, 43:7, 45:8, 65:17. In all cases, God is active in creating, not just at the beginning of time but continuously throughout history, particularly when there is an urgent need for his intervention. He is creating for his people today as much as he was in 600 BC during the Exile, or at the beginning of or before human history.

But what was God creating? We have to be very careful when interpreting the writings of the Old Testament, and particularly Genesis. A thorough-going interpretation begins with the original language(s) and finding in those languages the meaning of each word and phrase in terms of usage at the time of writing. We then need to find out how the authors would have thought about the subjects they were writing about, and in particular, how their culture differs from ours. This requires dedicated and painstaking study. It is well known that the creation accounts in the Bible are not unique in the ancient world. A number of similar accounts have been discovered in Babylon, such as the old Akkadian story, Enuma Elish[15]. By studying the cosmology of the Genesis 1, (that is, the way in which the Israelites regarded the origins and structure of the universe), John Walton, who is Professor of Old Testament at Wheaton College, argues that the cosmology of the period was *functional* and not *material* as we assume from our own cultural perspective[16]. In other words, the Israelites were primarily concerned with what functions things had in their ordered world. Because our thinking focuses on the material of what exists, we have to be careful not to impose our cultural viewpoint when interpreting the Biblical text. Walton states that 'The Israelites, along with everyone else in the ancient world, believed (instead) that every event was the act of deity—that every plant that grew, every baby born, every drop of rain and every climatic disaster was an act of God'[17]. There were no laws of nature or miracles[18]. The deity did not set things going and then leave them to their own devices, but 'he is thoroughly involved in the operations and functions of the world'[19]. This is a very different view of the cosmos than we have.

Genesis 1:1 states that 'In the beginning' God (the word 'Elohim' is used) created the heavens and the earth. Walton suggests that the Hebrew word for 'beginning' introduces the seven day period that follows in the text rather than a particular point in time[20]. At first, the earth was formless and empty, but there was water present, referred to as the 'deep', which had a 'surface' with darkness 'above' it. Although we can try to imagine what it was like, there was no light and so no image could be formed. It is tempting to think of the deep as primeval chaos, but the Hebrew word for 'deep': *těhôm*, is a good Canaanite word for the sea. (It is true however, that the Israelites did not have a very good feeling about the sea: for them it was a place of danger and threat.) Water is clearly a fundamental substance in conjunction with which everything else could be brought into existence. The important idea to note is that the water has as yet no function. Instead, the Spirit of God is 'hovering over the waters' (Genesis 1:2); that is, God was in a very real sense present with what he had begun to create. The Holy Spirit was pregnant with the creative acts of the Father and the Son. God would bring a functional creation into being with material water.

Then begin two periods of three 'days', when God incrementally added to his creation with an increasing complexity. Petersen (2005) comments that during the first three days God provided form to what was created, while during the second three days he focused on filling out the content[21]. What now is important is *how* God created. He called things into existence by proclaiming His Word; when God speaks, what is spoken comes into being. The New Testament writers insisted on the incarnation of the Word of God in the historical person of Jesus Christ (John 1:1-5). The early Christians in time came to acknowledge that Jesus Christ was in fact the second person of the triune God, and was active with the Father and the Spirit in creation. God's first words recorded in the Bible were 'Let there be light' and there was light (Genesis 1:3). Here, Walton regards God creating a period of light that alternates with a period of darkness. In other words, what God created was the sense of time passing as an endless sequence of periods of light and darkness. This is confirmed by God calling the light 'day' rather than 'light'. Walton comments that God 'had mixed time into the features of the cosmos that would serve the needs of the human beings he was going to place in its midst'[22].

God then reflected on what he had made and 'saw that the light was good'; that is, the light *functioned* properly. We are not told what the source of the light was, as the sun had yet to be created. And so evening followed by morning (remember, light had been created first, and evening was followed by night) announced the first day.

On the second day God spoke and created an expanse, called the 'Sky'. This solid expanse separated the waters above it and below on the earth. John Calvin in his *Commentary on Genesis* remarked how this is 'opposed to common sense, and quite incredible'[23]: but he was looking at Genesis from his cosmological point of view, which was different from the authors of Genesis, as it is from our modern scientific view-point. As Walton points out, the authors' cosmology was far more concerned with function rather than the material. Walton explains that they saw the solid sky (or firmament) providing a space for humankind below it, and a control on the water that came through it from above as precipitation[24]. In this sense, God created the function of the weather, which in particular moderated the amount of water coming from the cosmic waters above.

Interestingly, there is no record that God saw what He had just created on the second day and called it good. This could indicate that what He did that day was not complete, because on the third day God creates twice and each time saw that what he had created that day was good. In particular, on the third day God said 'Let the water under the sky (that is, below) be gathered to one place, and let dry ground appear' (Genesis 1:9). So it happened, and God called the dry ground 'land' and the gathered waters 'seas'. He 'saw that it was good'. But God had not finished his creative work on the third day. He said 'Let the land produce vegetation: seed bearing plants and trees on the land that bear fruit with seed in it, according to their various kinds' (Genesis 1:11). It was so, and God 'saw that it was good'. Walton suggests that these two acts of creation focused on the function of food production[25]. Like the weather, food and its production would turn out to be important for the arrival of human beings.

The form of creation had been determined; now it had to be filled with content. In Walton's terms, during the next three days God would create certain functionaries to fill out the functions created on the first three days[26]. On the fourth day God created the sun, the moon and the stars to populate the sky. Then on the fifth day God created living creatures to fill the water in the seas and birds to fly in the sky. The land would have to wait until the next day for its inhabitants. Again, each creature would have been made of a range of substances including water. Seeing that what he had created was good, God blessed the living creatures and the birds, and commanded the creatures to be fruitful and fill the water in the seas and similarly the birds to increase on the Earth.

Then on the sixth day God started by creating the wild and domestic animals and 'creatures that move along the ground'. After seeing that it was good, He called upon himself (in the plural) to 'make man in our image', with responsibility to rule over all the living creatures He had made. So man was created last, both male and female; for both aspects are needed for God's image to be formed: a notion that points to the importance of the man leaving his family and being joined to the woman so that they can become one flesh; see Genesis 2:24-25. In creating man in his own image, God defines man in terms of the relationship he has with him, in line with what God does and not necessarily in what God is;[27] see Drane (1986). Again, the man and woman are created with certain functions, especially procreation. This implies that God intends man to be an extension of his own personality, and to have a fundamental part to play as his representative in this world. Yet, as a created being, man's makeup is also consistent with the rest of life on our planet in that we bear a version of the DNA that is fundamental to all animals and plants[28]. However, we also bear the divine stamp, and, in this respect, we are separate from the animals. Man has been given consciousness and a moral conscience. This has important consequences for Christian apologetics today[29].

Like the creatures created on the fourth and fifth days, God blessed man, and said his creative word to them, authorizing them to be fruitful and increase in number, to fill the earth and subdue it. Every plant was given to the land-based creatures including man for food. Then God reviewed all that he had made, and it was not just good, but very good. There was a strong sense of rejoicing, or shalom, that Creation was complete (Proverbs 8:30-31), and God blessed the seventh day and made it holy, because He rested from all the work of creating He had done.

The seventh day account includes the repetition of the number seven three times. Eugene Petersen points out that the rhythm of the Seven Day account is in triplets: 1-2-3, 4-5-6 and 7-7-7[30]. What is more, the third day of each triplet has two acts of creation. Hebrew scholars see this theological account as being skillfully and rhythmically written in its original language, with each word in its proper place. Walton explains that Genesis 1 describes how the cosmos is given the functions to be God's temple, and therefore, his dwelling place. The world is God's headquarters, from where he runs the cosmos[31].

If Walton is right, and Genesis 1 is all about God creating functions and functionaries, then it leaves open how the material universe came into being. In this sense, the Bible is not in conflict with modern science (or the science of other periods in the past or future). This effectively defuses the current conflict between scientists and theologians, and opens up the way for better dialogue.

But there is yet another, separate account in the Bible of how God created the world. Whereas the account in Genesis 1 is about functions created during the Seven Days, the account in Chapter 2 is about the Garden of Eden as sacred space. We consider this second account in the next chapter.

For further reading:

Petersen E (2005) Creation and the gift of time, in Caring for Creation, ed S Tillett, Bible Reading Fellowship

Walton, J (2009) *The lost world of Genesis one.* IVP Academic

Resource questions:

Reflect on the place water has had as the resource from which the universe was made. What role does the hydrological cycle have in this account of creation?

Water is a basic component of life; but it is not life itself. What additional feature or function is needed for something to be alive?

Given that we are mostly made of water, what does this imply about our conservation and preservation of adequate fresh water resources?

What are the functions for which man was created? Which functions do you feel most affinity with, and why?

4

Water and Creation II

> 'We must treat water as if it were the most precious
> thing in the world, the most valuable natural resource.
> Be economical with water! Don't waste it! We still have
> time to do something about this problem before it is
> too late.'
>
> -Mikhail Gorbachev, President of Green Cross
> International, quoted in Peter Swanson's
> 'Water: The Drop of Life', 2001

Read Genesis 2:4-25

We come to the Garden of Eden account of creation in Genesis
Chapter 2. This was again written by a skilled story teller. The dry
land of the Earth is assumed along with the heavens. The oceans are
not mentioned; instead the focus is on the land which is watered by
streams coming up from below ground. At this stage there are no
plants because there is no rain. 'The LORD God formed the man from
the dust of the ground, and breathed into his nostrils the breath of
life, and the man became a living being' (Genesis 2:7). Water was
needed to complete the man's body. It could be that the water was
given when the LORD God breathed into the man's nostrils. The
implication is that as spirit is related to God's breath, so water is
related to spirit and therefore to the breath of God.

God had a task in mind for the man, namely to work the ground.
So God planted a garden of trees 'in the east', called Eden. Remember
that, according to the biblical account, there was no rain at this stage.
A river, which watered the garden, flowed out from Eden separating
into four headwaters: Pishon, Gihon, Tigris and Euphrates. The Tigris
and the Euphrates were well known not only today in modern Iraq but
also in the Old Testament, whereas Pishon and Gihon are difficult to
identify; some have suggested that they are rivers as far afield as the
Indus and the Nile[32]. The latest idea is that Gihon is the name given

The four rivers flowing from the Garden of Eden

to a spring in the Jehoshaphat valley, and Pishon is probably located in Mesopotamia near the modern day Persian Gulf[33]. Although small compared to the Tigris and Euphrates, each of these rivers can be regarded as one of the four rivers emerging from the Garden of Eden. The purpose of the four rivers was to make the physical source of life for the trees in Eden available for life elsewhere. Later in this book we will encounter the importance of rivers flowing from the temple and the throne of God in the new creation. In the meantime, man was to work the garden and to care for it. God said 'It is not good for the man to be alone. I will make a helper suitable for him' (Genesis 2:18). Subsequently God formed, again out of the ground, all the beasts of the field and the birds of the air to see if one of them would be suitable. None was found, so God resorted to making the woman, not from the ground like the other animals, but directly from one of the man's ribs, so that she is one with him. Again, water is involved in the creation of each creature. By naming the animals and the woman, Adam exerted his authority over them. But with the woman he also expressed his irrepressible joy at her companionship and ultimately his dependence on her.

What are we to make of the two separate accounts of creation and their relationship to modern science, especially to theories of evolution? Again, Walton has written concisely on this subject[34]. We have to be clear that the Genesis accounts are theological, and written by people within their own particular culture long before the appearance of modern science. They were written to express universal truths to explain the existence of our functional, rather than our material-physical, world, and particularly mankind. Therefore, there is no competition between the biblical account of creation and that of modern science. So far as the material universe is concerned, it is now generally accepted by scientists that it began with a 'big bang' from a point, as it were. But we cannot view the point from outside itself: we have to view it from within. Mathematicians have purported to be able, with their equations, to describe what happened back to just nanosecond (10^{-30} seconds) after the beginning of the bang, but they will not be able to sit back until they can explain the complete history going back to time zero. We await further insights into the physics of creation.

So far as life on earth is concerned, water is crucial. As a substance it is common throughout the universe: even Saturn's rings are largely made of ice crystals. It is thought that much of our water was brought by asteroids colliding with our planet when the solar system was forming. Darwin's theory of evolution appears to provide a satisfactory theory for the emergence and distribution of life on earth. There is still, however, no satisfactory scientific explanation for the appearance of new species: the time scale for the development of life on earth in all its details appears to have been remarkably brief. Similarly, life has become more complex as time has proceeded in apparent contradiction to the second law of thermodynamics (which states that in any closed system the amount of disorder increases in time, and therefore we would expect life to become simpler in time). John Polkinghorne[35] however, draws attention to the work of Prigogine, the Belgian 1977 Nobel Laureate in chemistry, who indicated that dissipative systems have an inflow of energy from their environment, while they 'export' their entropy[36]. This enables them to maintain a highly ordered internal pattern. In other words, Prigogine has shown that life can become more complex in time, and therefore the theory of evolution does not contradict the second law of thermodynamics.

In their optimism, many scientists say that they do not need God as a hypothesis to complete their explanation of the physical universe. Recently, Stephen Hawking declared that there is no need for God in order to initiate the 'big bang[37]'. What scientists like Stephen Hawking still cannot say from their scientific standpoint is *why* creation exists. Some scientists would say we do not need to ask this question. But it remains a theological issue. Some Christians, in opposition to the theories of the scientists, argue from Genesis 1 that God made the earth and the heavens in seven days each of 24 hours. There are a number of other possible interpretations. For example, one such interpretation is that Genesis 2:7 indicates that Adam was a two-stage creation. Sam Berry (2005) suggests that God created first pre-Adamic man, *Homo sapiens,* and then about 15,000 years ago, towards the end of the last ice-age He in-breathed His Spirit into a single individual Adam-Homo divinus[38]. This creation of *Homo divinus* then spread to all existing *Homo sapiens*. This adoptionist concept as an interpretation remains speculative, but it is consistent with the text of Genesis 2. What is at issue is that scientists still have an enormous amount to discover about how life formed on earth, and there are many renowned scientists who find no contradiction between their scientific work and the Biblical account of creation.

For further reading:

Berry R J (2005) *Rejection of the Creator,* in Caring for Creation, ed S Tillett, Bible Reading Fellowship

Biswas A K (1970) *History of Hydrology,* North Holland Publishing Company, Amsterdam

Drane J W (1990) *An Introduction to the Bible,* Lion Publishing

Petersen E (2005) *Creation and the gift of time,* in Caring for Creation, ed S Tillett, Bible Reading Fellowship

Polkinghorne J (1994) *Science and Christian Belief,* SPCK

Resource questions:

What do you think about our relationship to creation? Are we here to exploit it or to care for it?

Explore what others say about creation. What arguments are used by the creationists, and those using the argument of Intelligent Design?

Adam found it difficult to care for the earth beyond the Garden of Eden. Why was this?

Analyse your use of water this week. How can you make better use of water? Do you keep the tap running while you clean your teeth? Do you take a shower in preference to a bath? Does it make any difference?

5

Water and the Fall

American Indian proverb
quoted in 'Water Wasteland'
by David Zwick & Marcy Benstock, 1971

Read Genesis 3:1-24

When God created the earth, He deliberately ensured its sustainability by providing water for plant life and the creatures on land. Initially, water came up to the land surface, presumably from groundwater. Indeed, the river that came from Eden divided into four rivers: Pishon, Gihon, Tigris and Euphrates, which appear to have flowed in different directions to provide water for various parts of the earth. As mentioned above, whereas we are familiar today with the Tigris and Euphrates in present day Iraq, Pishon and Gihon are more difficult to identify. The reference to Cush in Genesis 2:9 could be coupled with Ethiopia and therefore the Nile. Although such a suggestion is tenuous it is worth considering in the context of what happened in Eden.

Adam and Eve were charged with caring for the garden, but it would appear this was not a full time occupation: they had time to relax and they had choice as to what to do with their time and will. On one occasion, they had a discussion with the serpent (who was also a creature created by God) about their relationship to God (whose name in Hebrew the Jews would not pronounce because it was so holy). This serpent represents cleverness and magical powers, and not necessarily satan. From what we know about temptation, getting involved in a discussion with the serpent was a grave mistake: Adam and Eve should have refused to answer its questions. The serpent persuaded Eve to eat fruit from the tree that God had forbidden to them. Eve gave some of the fruit to Adam, and after they had eaten the fruit, they both recognized that they were naked; in other words, they had lost their innocence when they gained knowledge of good and evil. In their subsequent embarrassing encounter with God, they learned the result of their disobedience. They were immediately banished from the Garden of Eden and access to the tree of life. The ground would no longer yield vigorous plant growth as previously. Adam would have to secure his food by hard work, the 'sweat of their brow' rather than living a gloriously fulfilled life, and Eve would have pain in bearing children while Adam would rule over her. Finally, they would die, after which they would return to the dust from which they had been formed. There would be an on-going conflict between Adam and the serpent, which would become a symbol of evil. Adam would crush the head of the serpent, who would strike his heel. The fall of Adam and Eve from their innocent relationship with God resulted in the whole of creation being affected. In particular, it would appear that the water from the four rivers, which was needed to provide water for plant growth and the well-being of animals, was less effective in sustaining the planet. The Fall would appear to have transformed the planet from being a luxurious idyllic environment which was fully self-sustaining into one which required effort on the part of man to make it produce necessary food. Indeed, the whole of creation has suffered because of the Fall; as Paul says in Romans 8:19-22, 'For the creation was subjected to frustration, not by its own choice, but by the will of the one who subjected it, in hope that the creation itself will be liberated from its bondage to decay and brought into the glorious freedom of the children of God. We know that the whole creation has been groaning as in the pains of childbirth right up to the present time'. Inanimate matter, as well as living things, is caught up in this experience.

An impression of the hydrological conditions after the Fall can be gained from the way in which modern day Egypt is supplied with water. Egypt is at the downstream end of the Nile; the country has very limited rainfall and its 79 million people are almost all dependent on water from the Nile. This river drains an area 75 times that of the Netherlands, it extends as far south as Rwanda and Burundi and water flows into it from ten countries. In 1998, in order to secure water supplies for 20 years ahead, the Egyptians built the High Aswan dam in the south of the country, to enclose Lake Nasser. This huge reservoir has the added benefit of removing the possibility of annual floods in the Nile delta, which used to occur before the dam was built. The flooding caused considerable damage, though it had the benefit of depositing a rich film of silt over the agricultural land, which regularly enhanced the fertility of the soil. The annual flooding has now been removed, and the farmers can irrigate their lands without fear of disruption, but the valuable deposition of silt has been lost. Because the water in the Nile is used time and again for domestic water supply, industry and agriculture, it is heavily polluted, especially the little water that actually reaches the Mediterranean. Inevitably, there is competition for the water resources from different sectors of the community. Indeed, the demand for fresh water throughout the Nile basin countries puts Egypt in a very vulnerable position as it has to respect the needs of these other countries. However, Egypt does have rights enshrined in international law which in effect gives it ownership of the rain that falls on those other countries.

The Fall has exposed the vulnerability of human kind. Having been given responsibility to care for creation as well as to rule over it, we have tended to regard the power we have been given as absolute. In order to maximize our benefits, we have chosen to exploit the earth's resources without regard to the future. Resources of oil, coal and gas are all non-renewable: once consumed they cannot be replaced. On the other hand, renewable resources such as timber and food require considerable effort for us to provide regularly adequate amounts for all. The approximately fixed annual quantity of fresh water is under threat due to the needs of the world's growing population, over-exploitation by communities upstream in a river catchment, the loss to the ocean of flood waters that cannot be stored, and unacceptable pollution due to untreated waste. Our inability to manage water in the atmosphere, along our coasts, and even in our major rivers has left us vulnerable to extreme events: floods, droughts and pollution. Water, we should remember, is not only an inanimate source of life on earth, it can also bring death. Human beings can drown in several centimetres of water, and cars can float away in water a half a metre deep. The devastating floods reported annually in different countries around the world, tsunamis destroying coastal communities as we have seen recently in the Indian Ocean and Japan, hurricane surges overtopping coastal defences as in New Orleans, and dam break waves washing out all before them, are symptoms that not all is as it should be in our world. Similarly, the absence of water is equally violent in denying the refreshment of the water integral to the survival of plants and animals and lost by dehydration. Global warming threatens to reduce rainfall from its present regime in some regions of the world, making the future of vegetation and animal life in those regions unsustainable; some species may survive, others will inevitably disappear.

What can we do? We share equally with each other in the Fall, but as Christians we claim to be born again of the Spirit of God. We therefore have an internal dynamic that opens our eyes, ears and hearts to what can be done, at least to minimize the damage done to the object of our care, namely all creation, if not to find ways of restoring it. We can do this with water by being more responsible in our use of it, using only what we need, taking care what waste we allow down the sink or drain, supporting Non-Government Organizations (NGOs) and other charities seeking to respond to the Millennium Development Goals, and praying that God's kingdom will come to renew the earth and minimize the corruption due to the human heart.

Resource questions:

What can you do today to contribute to the sustainability of water? Churches have sought to raise awareness of the needs of our planet through 'Green' initiatives. One emphasis is to focus specifically on how we use water. Have you made a permanent change to your lifestyle through a Green Group awareness campaign? Would it be worth repeating the initiative in the coming year?

6

Water and Climate Change

'Today we're seeing that climate change is about more than a few unseasonably mild winters or hot summers. It's about the chain of natural catastrophes and devastating weather patterns that global warming is beginning to set off around the world, the frequency and intensity of which are breaking records thousands of years old'.

Barack Obama, from a speech given on 3 April 2006 at 'Energy Independence and the Safety of our Planet' http://obamaspeeches.com/060-Energy-Independence-and-the-Safety-of-Our-Planet-Obama-Speech.htm

Read Genesis 9:8-17

The climate is changing continuously. So to talk about climate change as if it is a good or bad thing *per se* makes little sense. What is at issue is whether the climate is warming or cooling, how fast it is changing and what is the cause of it so that we can be prepared for what is likely to happen. In recent geological history, the climate has oscillated regularly between being warmer and colder than it is today. There are many interacting reasons for this oscillation, including cyclic variations in the sun's radiation, changes in the earth's orbit, cosmic rays, volcanic eruptions, dust and debris in the earth's path through space, human generated emissions of carbon dioxide and other greenhouse gases, etc. In order to trace the changes that have occurred, geologists have studied peat, pollen and spores from lake sediments, stalagmites and ice cores, as well as drawing conclusions from archaeology and geology. Plimer (2009), an Australian geologist, describes the history of climate change as deduced from observational data and from an extensive review of the scientific literature. What follows is a short description of climate change during Biblical times based on Plimer's analysis[39].

At the start of what is called the Holocene period, between 11,500 and 8,900 years ago the Earth experienced a warm period. Forests expanded rapidly, the tree line rose and glaciers retreated. The climate was not changing smoothly however; there were short, sharp cold periods interspersed with warmer times. The period 8,900 to 8,500 years ago was such a time of cooling that by 8,300 years ago there were large numbers of icebergs floating in the oceans[40]. Nomadic hunters and gatherers moved to lower latitudes, such as the Black Sea Basin, a quarter of which was then largely open grasslands at a level of 100 metres below the present sea level, surrounding two large fresh water lakes[41] (we discuss this in more detail later when exploring Noah's flood). The Northern Hemisphere warmed up again between 8,100 and 4,030 years ago, with the Arctic up to 3°C warmer than today. The Sahara was able to support groups of giraffes, hippopotami and elephants, and had its most lush period about 6,000 years ago[42]. The sea level was then about 2 metres higher than at present, until about 3,000 years ago (about 1000 BC) and Lake Chad was four times larger than the present day Lake Superior.

Civilizations rose and fell during the Palaeolithic and Neolithic periods (namely those periods in human history when stone and metal tools respectively were first used) due to extreme climate variability. In 4200 BC, prolonged drought caused the population in northern Mesopotamia to move south where the people irrigated the alluvial soils around the Euphrates[43]. The climate became warmer after 3800 BC, when the Sahara and the Arabian Peninsula still supported hunting, herding and some agriculture. This however, ended when there was another cool period between 3600 BC and 3300 BC. Such cooling eventually led to the desertification of the Sahara and the Arabian Peninsula by about 700 BC. There was extensive deforestation by herders and farmers throughout the Northern Hemisphere between 3500 BC and 3100 BC. This, together with over-grazing, lead to catastrophic erosion on steep slopes producing barren landscapes, such as found in modern day Greece[44]. The next warm period, to 2200 BC, led to further development of civilizations along the major rivers, including the Nile, though there were some devastating floods, in the Nile particularly[45].

There was global cooling by as much as 1.5°C beginning in 2200 BC. For 300 years there was global drought and famine, leading to the collapse of several civilizations. 1470 BC to 1300 BC was a warm period followed by further global cooling between 1300 BC and 500 BC[46]. It was about 1280 BC that the Israelites were escaping from Egypt *en route* to the Promised Land. The Hittite empire went into decline in 1200 BC and disappeared soon after; Egypt also declined, and Babylon and Assyria were weak for a hundred years after 1100 BC. There were again massive floods in the Nile around 800 BC, and the climate cooled between 750 BC and 450 BC, which was when the Assyrian, the Babylonian and subsequently the Persian empires were at their zenith.

The climate started warming again in about 250 BC (the Roman Warming)[47]. The Greek and Roman empires benefited from the change as temperatures became 2°C to 6°C warmer than today. Certainly by 300 AD it was considerably warmer. North Africa became the breadbasket for the empire. Sudden cooling took place in 535 AD and 536 AD which heralded the Dark Ages; it became cold; there was war, plague, famine and disaster[48]. The Black Sea froze in 800, 801 and 829 AD, and even the Nile had ice on it. Subsequently, the Medieval Warming occurred between 900 AD and 1280 AD[49], which was the period when Muslim imperialism and culture were at their peak. But during the 23 years after 1280 AD, major climate cooling took place, and the Little Ice Age started. There were few sunspots and little solar activity, and there was increased cloudiness. The periods 1450-1540 AD (called the Spörer Minimum after the man who studied it), 1645-1715 AD (called the Maunder Minimum) and 1795-1825 AD (referred to as the Dalton Minimum) were particularly cold[50]. Drastic changes in temperature, storm intensity and precipitation during these periods had profound effects on human society, affecting economic development, inducing famines and influencing culture; see, for example, the paintings by the Dutch painter Breugel (1525-1569) of people skating on ice. The climate has been warming again during the period 1850 AD to the present. Plimer claims that we are entering another period of global cooling[51]. The International Panel on Climate Change (IPCC) states that the climate is becoming warmer still due to man-induced change through the production of greenhouse gases such as carbon dioxide, which has been accumulating since the start of the Industrial Revolution, and is at levels of concentration that are far larger than have occurred during the last hundred million years; see IPCC (2007)[52]. The IPCC report of September 2013 confirms that the committee is 95% confident that global warming is being caused by human activity through the large concentration of carbon dioxide continuously being pumped into the atmosphere, particularly by industry[53]. One anomaly however, is the apparent constancy of the average global surface temperature between 1995 and 2013[54]. Despite this, the IPCC says there is an urgent need for political action between the industrial nations to reduce carbon dioxide in the atmosphere by reducing emissions, and to explore solutions for the growing crises that will emerge in different parts of the world as sea levels rise. The consequences for the global weather pattern appear to be the increasing frequency of presently rarer events, such as damaging hurricanes, extreme monsoons and flash floods[55].

What we need to observe however, is that the climate has always been changing, and will continue to do so. Obviously, it has affected the history of Israel and the early Christian church. When Abraham left Ur sometime after 2000 BC he was possibly persuaded to travel west by the harder conditions created by climate cooling. Similarly, the Israelites escaped from Egypt during another cooling period. Interestingly, the period from the reign of Solomon (circa 972-932 BC) through to the Exile (circa 582-538 BC) was also one in which there was climate cooling. It appears that the major events in the history of Israel are coupled with cooling of the climate rather than warming. In other words, climate cooling brought about a degree of instability in the international picture which resolved into Israel and Judah having to endure some tough experiences, something that has been sustained down to modern times.

Generally the climate has changed gradually over centuries. However, there are possible events that can introduce catastrophic change within a few years. One such event is the impact of a sizeable asteroid, such as is thought to have happened causing the extinction of the dinosaurs. Another event is a violent volcanic eruption which introduces large amounts of dust into the atmosphere. This can lead to significant cooling of the planet for a few years, leading to loss of vegetation and poor agriculture, as happened with the eruption of the Tambora volcano in Indonesia in April 1815[56]. Yet another event could be the switching off of the Gulf Stream, the major ocean current that is responsible for keeping Europe free of ice during the winter. The switching off of the Gulf Stream could be induced by the collapse of the Greenland ice-cap into the sea.

Whatever the future holds, we do well to acknowledge that God, our heavenly Father, is not only aware of the present state of the climate but is working with the forces on the planet and in the universe to bring about his purposes. We can see this in the major climate events of recent geological history. We do well to keep informed about climate change and to encourage others to be so too.

For further reading

Plimer I (2009) *Heaven and Earth: Global warming, the missing science.* Taylor Trade Publishing

IPCC (2007) 4th *Assessment Report* http://www.ipcc.ch/

IPCC (2013) 5th *Assessment Report* http://www.ipcc-wg2.gov/SREX/

Resource questions:

National governments are very concerned about the possibility of climate warming being due to the generation by industry and transport of carbon dioxide and to the destruction of the rain forests, which means there are less trees to remove this gas from the atmosphere. What can we do as a church to contribute to this debate and the promotion of appropriate actions?

What are the likely consequences to your country of global warming? How could you support your government or your local council in getting Christians and the Churches to help counter the adverse effects of global warming through particular initiatives?

7

The Flood

*'Whenever the rainbow appears in the clouds, I will see
it and remember the everlasting covenant between God
and all living creatures of every kind on the earth.'*

Genesis 9:16.

Read Genesis 6:1-8:22

A careful reading of the creation accounts in Genesis reveals that all
was not as we might expect it to be after the expulsion of Adam and
Eve from the Garden of Eden. Rivers flowed from the garden to water
the earth, but there was no rain; in other words, the hydrological cycle
as we know it today was apparently not yet functioning. This biblical
mystery is resolved by the account of Noah's flood.

The people had begun to spread out over the Earth, but their
awareness and relationship to God was lukewarm. Indeed, men
were so bent on following their own thoughts and feelings that
they became corrupt and violent. God concluded that he needed to
'wipe mankind, whom I have created, from the face of the earth—
men and animals, and creatures that move along the ground, and
birds of the air' (Genesis 6:7). God grieved that he had made them
in the first place. But in order that there would be continuity (or a
'remnant') of each species of animal, reptile and bird, God identified
one righteous man, Noah, and instructed him to build an ark to convey
two of every kind of creature (but including seven of certain 'clean'
animals), together with his own family, safely through a devastating
flood. Noah was another sort of Adam. His father, Lamech, named him
Noah because the word in Hebrew sounds like 'comfort', and Lamech
sought comfort from his son for the pain and toil he experienced
because of the curse on the ground due to Adam's sin. Noah was the
true remnant of humankind through whom God would address the
darkness in the heart of man and bring about His purposes. The flood
would be so severe that it would destroy every living creature left

on the earth. Noah did precisely what God asked him to do, strange as it seemed to those around him: he then loaded the ark with the living creatures, his family and the necessary provisions. Seven days later 'the springs of the great deep burst forth, and the floodgates of the heavens were opened' (Genesis 7:11). At long last it rained, but this was no preliminary shower, it rained continuously for forty days and forty nights. The water that God had separated above the heavens on the first day of creation was finally falling to earth. It was a resolution of that first day, indeed of the whole of creation; each month of the flood reflected a day of the first creation narrative. There was so much water that the tops of the mountains were covered by about seven metres. The waters continued for a hundred and fifty days. No person or land-based living creature survived other than Noah and those with him in the ark.

If we assume that the flood covered the whole earth, it is difficult to resolve where all the water came from—or where it went. Even if all the ice caps, sheets and glaciers melted, the sea level would only rise several tens of metres[57], which is not enough to cover even the low hills on all the continents. In line with other similar accounts of a great flood, it is possible that the phenomenon could refer to a much more limited but none-the-less devastating event. Critics of the biblical account identify two separate sources (which are labeled 'J' and 'P' and are well established in other parts of the Pentateuch, that is, the five books of Moses) that have been merged into a single text. The search for the origin of these sources has led scholars to similar flood stories in other contemporary countries and regions. The Epic of Gilgamesh is one such source, which was discovered in the Library of Assurbanipal in Nineveh, and dates from the middle of the seventh century BC[58]. It records a massive flood occurring in Mesopotamia. Another source is the history of Babylon by Berossus written in 300 BC[59]. Archaeological evidence reveals that the Tigris and Euphrates region was subject to large floods in Ur and Kish. This has always been a popular story with the Hebrews and the Jews, who appear to have taken it from Mesopotamian sources and added their own details, especially regarding God's attitude to human kind. What the authors of the biblical account conclude is that God regarded Noah as another Adam, with whom he entered into a covenant relationship. Note that the inconsistencies in the account, such as the numbers of animals of each species entering the ark help to enhance its authenticity rather than limit it.

As an alternative to the location of the flood in Ur and Kish, Plimer (2009), (the Australian geologist with specific views about climate change and global warming in Chapter 3), proposes the sudden breakthrough of the Mediterranean at the Bosphorus to the area covered by the present day Black Sea as the origin of Noah's flood[60]. He points out that the 160,000 sq km basin had a warmer, wetter climate than the Anatolian highlands in the exceptionally cold period 8,500 to 8,000 yars ago, and as a consequence people migrated to the basin. A quarter of the area was flat and covered with grasslands, and there were two large freshwater lakes in the middle. During the last glacial period, the Sakarya River drained the basin into the Mediterranean Sea via the Gulf of Izmet and the Sea of Marmara. With the rise in sea level of 120 metres after the glacial period, the Marmara Sea level was about 100 metres higher than the floor of the Black Sea Basin. There was a ground movement (that is, an earthquake) along the North Anatolian Fault in about 5600 BC', which led to water breaking through at the Bosphorus from the Sea of Marmara into the Black Sea. This inflow was completed in about two years. Of course, this would not explain Mount Ararat as the resting place of the ark. Some scholars propose that in line with the suggestion that the flood was located in Mesopotamia, the reference to Ararat in the Bible could be to the kingdom of Urartu, which included the highlands of Northern Mesopotamia, and included the region of Eastern Turkey around Lake Van.

In the biblical account, the recovery from the flood was slow. As the waters receded, the ark (as already mentioned above) came to rest on Mount Ararat. This resting of the ark can be viewed as the parallel of God resting on the seventh day of creation. In other words, it reinforces the notion of the flood leading to a recreation of the Earth. But Noah wanted to find out whether the exposed land was capable of supporting life. He sent a raven to look for some indication of the present state of the land, but the raven did not come back. Seven days later, Noah sent out a dove, which returned as it found no place to rest. But the second time the dove was sent out, it returned with a freshly plucked olive leaf. The dove was sent out a third time, and it

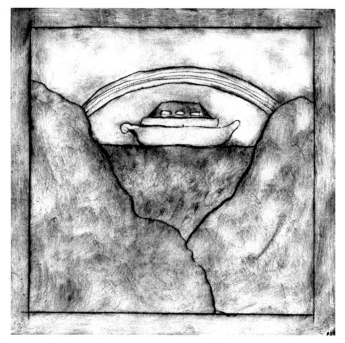

Noah's ark and the flood

did not return. The wind, like the Spirit of God over the face of the primeval waters, blew to dry up the waters of the flood. Once the Earth was completely dry, Noah was instructed to come out of the ark and to get on with life. Noah built an altar to God and sacrificed some clean animals and birds. God smelled the pleasing aroma, and resolved never to curse the ground again because of humankind, even though he acknowledged every inclination of each person's heart is evil from childhood. He also determined never to destroy all living things by such a flood. 'As long as the Earth endures, seedtime and harvest, cold and heat, summer and winter, day and night will never cease' (Genesis 8:22). But God still had a deeper plan in mind to redeem the people he had created. (Later chapters in this book explore the outworking of God's plan[61].)

God blessed Noah and his sons and repeated the command he had given Adam to 'be fruitful and increase in number and fill the Earth' (Genesis 9:1). The first man and woman had been given just the plants for food; now God gave Noah every clean living creature to eat, with one condition: he must not eat meat with the lifeblood still in it. Then God established his first covenant with Noah and his sons, their descendants, and every living creature on Earth. Never again would all life be destroyed by a flood, and the rainbow was to be the sign of this covenant. A rainbow depends on the presence of water drops in the atmosphere (not just vapour) with the light of the sun shining through them. Each drop acts to refract the 'white' light from the sun, breaking it down into its component wave lengths, and so generating the well-known arc of recognizable colours: red, orange, yellow, green, blue, indigo and violet. Water drops can be produced by the spray from a waterfall, so that a rainbow can be seen in the spray with the sun shining on it, but of crucial importance for Noah was the sun shining in the sky simultaneously when it was raining. The rain would always carry with it the threat of global flooding, but the appearance of the rainbow would remind Noah—and us—of God's covenant which reassures us that the threat will not be fulfilled again.

Like many of the incidents in Genesis, this account of the rainbow seems to have something of the characteristics of the 'Just-so' stories of Rudyard Kipling. Indeed, we should not be surprised to find elements of the biblical account in traditions of other cultures, for as Paul explains in Romans 1:20, everybody is aware of God's power and divinity because they are clearly seen in what has been made. Therefore the myths of different peoples and cultures convey elements of truth, and according to C S Lewis it is Christianity that provides the overall structure to make sense of these myths. Undoubtedly these myths formed an integral part of a strong oral tradition in which the stories were repeated word-for-word down the generations.

There are some interesting conclusions embedded in the biblical text about Noah's flood. Peter Harris (2005) points out that by taking male and female of each animal and bird species into the ark, God ensured that Noah became their carer, and therefore we by default are also responsible for them[62]. Indeed, the terms of the covenant are stated as much to the benefit of all other creatures as for Noah; see Genesis 9:8-17. Therefore God's covenant love is as much with the living creatures on earth as it is with humankind. The conclusion is far-reaching. It means that we have every responsibility to preserve and maintain each species. This flies in the face of the way we exploit creation, ripping down tropical and other forests and destroying complete species as we go, let alone taking advantage of the human poor to increase wealth. The Christian church needs to stand up and be counted in respect of preserving endangered species, lobbying to maintain and rehabilitate forests, defending the poor and oppressed and determining what we can do personally in our own local area to put these principles into practice.

For further reading:

Harris P (2005) *God's covenant with the earth*, in Caring for Creation, ed S Tillett, Bible Reading Fellowship

Plimer I (2009) *Heaven and Earth: Global warming, the missing science.* Taylor Trade Publishing

Resource questions:

What are the implications for climate change of the covenant God made with Noah?

How can we as individuals or a church adhere to the covenant God made with Noah?

Refer to p 150 of Alister McGrath's biography of C S Lewis: 'A Life', where Alister writes that 'for Tolkien, a myth is a story that conveys 'fundamental things'—in other words, that tries to tell us about the deeper structure of things. . . . . Myths offer a fragment of the truth, not its totality. . . . . For Tolkien, grasping Christianity's *meaningfulness* took precedence over its truth. It provided the total picture, unifying and transcending these fragmentary and imperfect insights.' C S Lewis took on board Tolkien's view and developed it in his writings. What does this mean for our appreciation of non-Christian myths?

Read pp 218-229 on 'Mere Christianity' in the biography of C S Lewis by Alister McGrath, 'A Life', and explore Lewis' proposal that 'there is a notional, transdenominational form of Christianity, which is to be cherished and used as a basis of Christian apologetics; yet the business of becoming or being a Christian requires commitment to a specific form of this basic Christianity'. What does this mean for the way in which you live your life?

8

Water in arid areas

'When the well is dry, we know the worth of water.'
Benjamin Franklin, Poor Richard's Almanac, 1746

Read Genesis 26:14-17

The notion of a super flood that affected the whole planet is difficult to appreciate considering the arid nature of the Middle East during Biblical history. The trade routes in Abraham's day linked the major centres of civilization on the Nile and the Tigris-Euphrates valleys. They traversed the Fertile Crescent, a reasonably fertile arc of land between Egypt and Nineveh that was watered either through abstraction from the rivers Euphrates, Tigris or Jordan or through moderate annual rainfall. It was along these routes that Abraham set out to the land God had promised him and his descendants. Traveling with flocks and herds, the actual route that Abraham took was dictated by water sources, and particularly oases and wells. Certainly, after he reached his destination, he settled where there were springs or wells that his men or others had dug. Where there were no sources of fresh water on the ground surface, such as in natural springs, streams, rivers or lakes, people had to resort to water underground.

The location of wells partly depended on the area where people chose to live. Their positions were dictated by some knowledge of where water was accessible not too far below ground. A well was lined as it was dug to prevent it collapsing inwards. The aim was to reach an aquifer, that is, water that saturated the rock formation at a certain depth, filling the pores or voids in the strata. An 'aquifer' consists of something like an underground lake or river; the water in the aquifer can be stationary or moving very slowly as it seeps through the rock strata. The aquifer usually has a definite surface below which the rock formation is saturated like a sponge. The well would be dug down somewhat deeper than the aquifer surface so that water would pond in the bottom of the well, largely from the aquifer. Water was withdrawn from the well, probably using a simple pulley system to raise and lower a container maybe made of animal skin; water would seep into the well from the surrounding aquifer to replace what was withdrawn. If it took a long time to replenish the well, the water tended to grow 'stale', possibly with algae growing in it over time, whereas if the water was replenished quickly, it would be like a spring existing at the bottom of the well. Then the well would yield a good amount of fresh water, or 'living water' as referred to by Jesus in his conversation with the Samaritan woman at the well (John 4:11). An advantage of using water from such a well was that, provided it was covered and rubbish was not thrown in, the well was generally free from human contamination. This was in contrast to many rivers which regularly received human and animal waste. The water from some wells, however, contained particular toxic chemicals, such as arsenic and fluoride that were dissolved from the rocks. Such wells became known for their unsatisfactory water, and were avoided or filled in.

Digging a well took time and effort, so each well was valuable in its own right. The well normally belonged to the person who dug it, but the importance of water for survival in an arid land often meant that wells were made available to all. Obviously, an area of land and its wells could support only a limited number of people and animals. Access to wells often led to conflicts, especially between shepherds and herders who needed water for their flocks, as when Isaac and his shepherds came into conflict with Abimelech, king of the Philistines, over the watering of their flocks and herds (Genesis 26:14-16). This conflict continued for some time, though it did not result in open warfare. During times of war, those defending an area of land would often fill in the local wells to prevent the aggressors having access to their water, thereby making it difficult for them to sustain their presence in the land. The Philistines did this to some wells which Abraham had dug (Genesis 26:15). After the conflict was over the wells had to be re-dug (Genesis 26:17).

Because wells were important to the local communities that were dependent on them, they became ideal meeting places. Abraham's servant met Rebekah, the grand-daughter of Abraham's brother Nahor, at a well in the town in Ur where Nahor lived, and she subsequently returned with the servant to become Isaac's wife

(Genesis 24). In a similar story, Jacob, Isaac's son, met Rachel, the daughter of Laban, Nahor's grandson, at the well in Paddan Aram.

New wells are now drilled (or bored) with machinery rather than dug by hand. Wells continue to be important as sources of fresh water in many countries today. They are also places where animals are watered, clothes are washed, and water is collected. Most of the collection and carrying of water is done by women and girls, many of whom walk long distances every day to provide sufficient water for their families. The need to fetch water prevents many young girls attending school regularly, and they remain uneducated the whole of their lives. One of the Millennium Development Goals is, by 2015, to provide fresh water for at least half of the 1.2 billion people who were without it in 2005[63]. This target applies to both urban and rural areas. Wells are expensive to dig (or drill with machinery) and need to be sited with care to ensure that they yield sufficient water when they are dug; similarly, wells should not be too close together or too far from where people live, and they should be at least 100m away from the nearest pit latrine. Local people who benefit from the wells also need to be encouraged to take responsibility for maintaining them properly. A number of Non-Government Organizations (NGOs) and charities have accepted this challenge and are endeavouring to meet the need of creating the supply of fresh water to people without it. Churches and other faith communities can encourage larger communities to take ownership of and responsibility for their water supply and wastewater disposal. One particular organization focusing on encouraging community responsibility for water supply is a Dutch registered NGO called the Healthy Vine Trust. This trust works closely with the Church in Uganda in the Diocese of Luweero to drill new boreholes to provide wells in the parish of Sekamuli, as part of a more wide ranging project to reduce malaria in the region[64]. In addition there are a number of important charities, such as Water Aid and Aqua For All, working to bring freshwater and healthy sanitation to peoples and communities world-wide.

Competition for fresh water will increase in the future as the human population of the world increases; new sources of water will need to be explored. A project in Libya, initiated well before the revolution in 2011, is particularly interesting; the government decided to exploit the huge underground water resource below the

Sahara desert by pumping water thousands of kilometers to the Mediterranean coast where most people live. In past millennia, this underground resource has been used with great care, because the water is not renewable, at least in the near geological time scale. Libya seeks a short term gain from the resource, in the hope that some other water source, such as desalination, will become available in the future. It is interesting to reflect that sea water offers us the hope of solving two of our greatest crises: our needs for energy and fresh water. In the case of energy, small amounts of deuterium in sea water can be extracted for nuclear fusion with tritium[65], while fresh water can be obtained by evaporating sea water and then condensing it back into a liquid, though this takes a great deal of energy[66].

For some NGOs working with water in developing countries, see The Healthy Vine Trust, http://www.healthy-vine.org/water.html

Water Aid, http://www.wateraid.org/uk/

Aqua For All http://www.aquaforall.nl/uk/about-aqua-for-all/ organization

Tearfund, http://www.tearfund.org/About+us/What+we+do/ Improving+basic+services.htm

A Rocha, http://www.arocha.org/int-en/index.html

These web sites were accessible when the book was published. If for some reason they become inaccessible, please use a search engine to locate the appropriate web sites.

Resource questions:

How can we be more informed about the Millennium Development Goals (MDGs)? What can we do as a church to assist the achievement of these goals?

See if you can find out how much it costs to drill a well in a developing country and to maintain it for water supply. How can you support agencies seeking to bring fresh water to those without it?

9

Water and the Exodus

'Nothing in the world is more flexible and yielding than water. Yet when it attacks the firm and the strong, none can withstand it, because they have no way to change it. So the flexible overcome the adamant, the yielding overcome the forceful. Everyone knows this, but no one can do it.'
Lao Tzu quotes (Chinese taoist Philosopher, founder of Taoism, wrote 'Tao Te Ching'. 600 BC-531 BC)
See http://thinkexist.com/quotes/lao_tzu/

Read Exodus 14:5-31

Morecambe Bay in England is notorious for its extensive mud flats which are exposed at low tide. They invite people to dig for shell fish, and particularly cockles. But the incoming tide travels at the speed of a 'good horse' and can cut off unwary cockle pickers. This is what happened on the evening of 5 February 2004, when a group of 21 Chinese illegal immigrants died, despite attempts to warn them of the danger they were in. It is no surprise that the sea contains many hidden dangers for the unwary and inexperienced. In general the Old Testament talks little about experiences with the sea, for the patriarchs and the people of Israel were deeply suspicious of it. International trade by sea was typically carried out by other groups and nations. Biblical references to the sea are almost always couched in terms of chaos and potential, if not actual, disaster.

The sea was probably far from Jacob's mind when a serious drought affected the whole of the Middle East. At the invitation of his son, Joseph, Jacob took his family to live in Egypt. Years before, Joseph had been sold by his brothers, and became a slave in Egypt, but he had prospered to become second in command in Pharaoh's Egypt. Joseph had secured the survival of Egypt as a nation by successfully forecasting and managing seven years of plenty when the River Nile regularly supplied its annual flood, followed by seven years of famine when the Nile floods and the associated crops failed. He made it possible for his father and the whole family to live in Egypt where food from the years of plenty had been stored. Over a subsequent period of 400 years, the 70 members of Jacob's family became a large group of 12 tribes, called the Israelites (or Hebrews) after Jacob (also called Israel).

As with any immigrant people that does not integrate with the host population, the Israelites came to pose a significant threat to the Egyptians, who took action to limit the threat by forcing them into slave labour to build the cities of Pithom and Rameses. In the darkness of their suffering, the Israelites cried out to the God of their ancestors for deliverance. In response to their prayers, God took the initiative to call Moses. When he was a baby, Moses had been placed in a waterproof basket and hidden in bull-rushes at the bank of the River Nile by his Hebrew mother. He had been found and adopted by Pharaoh's daughter, who brought him up in the Egyptian royal family. As an adult, Moses had to flee from Egypt, after it became known that he had killed an Egyptian for abusing a Hebrew slave. He spent 40 years as a shepherd in the land of Midian, (looking after the sheep of his father-in-law) and there he encountered the LORD at the burning bush; see Exodus 3:1-4:17. God told Moses to return to Egypt and to negotiate with, or rather, tell Pharaoh to let God's people go to worship him in the wilderness. Their ultimate objective was to return to the land God had promised centuries before to Abraham. So Moses, with his brother Aaron, met with Pharaoh on a number of occasions to persuade him to let the Hebrews leave Egypt. These meetings turned into a major confrontation between God and Pharaoh. Pharaoh was not prepared to give up the benefit of the Hebrews' slave labour, and in response to Moses' demands made the conditions of their slavery worse, despite Moses calling down a number of plagues that afflicted the land and the Egyptian people. The Nile water was turned to blood, then frogs, flies, boils and hail infested everything, locusts ravaged the land, there was deep darkness, and even the first born of men and animals were destroyed. This last plague finally broke Pharaoh's resistance, and he let the people go into the wilderness to worship their God. But once he had made the decision and the Hebrews had left, he regretted it and wanted to round them up and return them to their slavery.

But the Hebrews were on their way. They journeyed east with the objective of returning to the land of Canaan where their ancestors Abraham, Isaac and Jacob had lived. Such was the dramatic nature of their flight from Egypt that God was present with them in a very real way: he went before them in the pillar of cloud by day to lead them and behind them in a pillar of fire at night to protect them. In the meantime, the Egyptian army with their chariots and weapons of war gave pursuit. Humanly speaking, the situation of the Israelites was dire. It became even more serious when the escaping Hebrews came up against a stretch of water linked to the sea, which was so deep they could not cross it. They were trapped. What could they do? Turn and surrender to their erstwhile captors? But Moses had sufficient experience of God's care for His people to know that God was capable of sorting out the situation, even if he, Moses, did not know how it would be done. His responsibility was simply to do what God told him to do. And this he did. God told Moses to hold out his staff over the water in front of him. As he did so the waters parted with walls of water on the left and the right so that the people could walk straight ahead on dry ground where once the water had been.

It must have taken time for the people to walk to the other side of the Red Sea, looking in amazement from side to side at the water piled up, wondering whether it would return to swamp and drown them, or whether the Egyptian army would catch up with them. But they continued walking. Once there, they were aware that the Egyptian army was not far behind. The chariots were charging through the gap between the walls of water, precisely where the escaping slaves had walked a short while before. Were they to be caught again as freedom beckoned? Then Moses was commanded by God to hold his staff out over the water again, and as he did so the waters returned, the Egyptian chariot wheels foundered in the soft sand and mud, and the whole Egyptian army was swamped and drowned. The threat was over once and for all. The Hebrews were free; they had miraculously crossed the boundary formed by the Red Sea; they were no longer slaves of the Egyptians. The water had given life to the Israelites and brought death to the Egyptians.

Moses and the Israelites celebrated their escape and the destruction of the Egyptian army with a song (Exodus 15:1-18). In it,

Crossing the Red Sea

they emphasized that it was God alone who shattered the enemy, and that the nations would hear and tremble. In particular, the people of Canaan would melt away with fear as God planted his people in the land. This act of recalling the amazing work of God in rescuing the Israelites from Egyptian slavery was enshrined in stories and songs. These were to be rooted and grounded in the collective memory of God's chosen people, so that it would be as if each generation had personally experienced what God had done. It is therefore not surprising that the Psalms make frequent reference to the Exodus; for example, see Psalms 66:6 and 106:9. Isaiah also refers to the Exodus, as in Isaiah 10:26 and 44:47. Even the New Testament refers to the Exodus account as in Acts 13:17 and 1 Corinthians 10:1. The significance of the Exodus is that it marked the beginning of Israel as a nation. It was a baptismal experience for the people who escaped; an initiation into being set apart as the people of God. Water had played a vital part in that initiation, as it does for every Christian through their baptism in the name of the Trinity: the Father, the Son and the Holy Spirit.

What is more, water formed the context of one of the most amazing miracles in human history. The Exodus was such a crucial event in the formation of the nation of Israel that God wanted his people to remind themselves of the miracle(s) he had performed on their behalf. He instructed then to recall these seminal events in their history, to make them regular topics of conversation, and in particular to pass on the knowledge and experience of these events to their children, that is, to the next generation. Psalm 22:3 states that ' . . . you are enthroned as the Holy One: you are the praise of Israel'. An alternative reading of this verse says that ' . . . . you are holy, enthroned on the praises of Israel'. Indeed, to have a living relationship with God we need to join together in worship; whether singing contemporary songs with a real sense of rhythm, or being immersed in the beauty of choral evensong in one of the great cathedrals of the Anglican Communion. The retelling of the old stories of what God has done, especially for his people, are to be repeated again and again so that they are a true part of our (Christian) culture, as well as that of the Jews. The crossing of the Red Sea is as much for modern day Christians as it was for the early Israelites.

As a postscript to this chapter we should remember that the significance and importance of the exodus from Egypt have attracted the attention of the scholars who seek to establish what actually happened, separating myth from reality. For example, there has been much debate between scholars over the route that the Israelites took in their escape from Egypt. Most of the recent evidence suggests that the route did not actually take place through the Red Sea as is traditionally believed, but to the north through a region crossed by the present Suez Canal[67]. For 'Red Sea' we should read 'sea of reeds'; that is, a marshy area that would have in effect been a wetland with reeds and lakes. Presumably the lakes would have been deep enough to cause drowning of those caught without assistance. The miracle could have been due to the timing of a strong wind that blew across the path of the escaping slaves so that water was piled up to at least one side, producing a comparatively dry path through the 'sea'. Once the wind stopped with the Egyptian army in pursuit the path through the 'sea' was no longer available, and the pursuers were no more. Like the tide in Morecambe Bay, the water would have returned so fast that the Egyptians could not escape back the way they had come.

They could not even reach the escaping slaves. The threat from Egypt was over once and for all.

Resource questions:

What boundaries have you crossed where your life has been radically changed?

What are the stories that you recall are formative for you and your family?

What stories do we tell ourselves in our Christian communities in order to recall God's grace and love?

In what ways was water involved in the plagues that God brought about through Moses and Aaron in Egypt? How does the notion of plagues recur in the Bible?

10

Water from the rock

'He split the rocks in the desert,
And gave them water as abundant as the seas;
He brought streams out of a rocky crag,
And made water flow down like rivers.
But they continued to sin against him,
Rebelling in the desert against the Most High.
They wilfully put God to the test
By demanding the food they craved.'

Psalm 78:15-18

Read Numbers 20:1-13

The Exodus occurred during a period when there were reasonable water resources in the land between Egypt and Canaan. But the large number of Israelites with their flocks and herds ensured that there would regularly be stress on those resources. A shortage of water puts people on edge, especially when they think that they will not have sufficient to stay alive. In such circumstances they can readily come into conflict with others, especially if they perceive them to be (partly) responsible for their predicament. There have been many local conflicts over water, though fortunately to date there have been few, if any, wars between neighbouring states.

Conflict between the Israelites and their leader Moses was never far below the surface. Having followed him from Egypt, they were buoyed by the amazing things that had happened to them, including crossing the Red Sea and seeing the Egyptian army routed and drowned when the water returned to cover the path they had taken. God was a God who could supply their needs. The question was: would God do so in every circumstance? They were now in a position to find an answer to this question.

Being isolated in the wilderness between Egypt and the Promised Land, water and food began to be in short supply. The tension increased as the peoples' mentality, profoundly influenced by their slavery in Egypt, was one of looking to Moses (*in loco parentis* instead of the Egyptian taskmasters) to provide their needs. What was Moses doing in bringing them to such a place where there was no fresh water? Had God forgotten them? Had he brought them out into the desert simply to destroy them as he did the Egyptians? They could only survive for a few days without water. Their children would be the first to suffer. What was Moses going to do about it? How was God going to get them out of this mess? So they grumbled, firm in their slave mentality, unable to find the solution for themselves. When people grumble to each other, they reinforce their mutual discontent. The grumbling increased until the Israelites were ready to stone Moses, whom they regarded as the visible sign of God's presence; in this way they could take out their frustration on God. What they were doing was testing whether God was among them or not; see Exodus 17:1-7.

There was no water for the community and the people were fomenting rebellion against Moses and Aaron. There appear two accounts of what happened next; see Exodus 17:1-7 and Numbers 20:1-13. The first was when the community camped at Rephidim. The people drove Moses to distraction over their demand for water. In response to his appeal, God told Moses to take some elders with him to the rock at Horeb and to strike the rock with the staff he had used when striking the Nile. Water came out of the rock and the people drank. The elders were witnesses of the event. Moses gave the place two names based on the people's actions: Massah, or 'testing', and Meribah, or 'quarreling'. It appears from the biblical record that the people came to Horeb a second time with the same attitude as before; see Numbers 20:1-13. Care is needed however, as the two major sources, 'J' and 'P'[68], identified for the original text used by a supposed editor, can repeat material or even be in conflict with each other. However, we assume that the account in Numbers provides a second encounter with a rock when the people desperately needed water. Again the people were ready to stone Moses, who became exasperated with them. On appealing to God, he was told to go and *speak* to the rock nearby and water would flow from it. Whether or not he remembered the first occasion he had been there, he collected his staff and took Aaron (rather than the elders) with him to the rock. The people followed them, muttering threats and ready to pick up the nearest stones. Once at the rock, instead of doing what God commanded him, namely to speak to the rock, commanding water to comefrom it in God's name, Moses asked the people if they wanted him to bring water from the rock. He does not appear even to have waited for an answer, but struck the rock with his staff. Water gushed from the rock, so much so that it again met the needs of all the people. But this time Moses had disobeyed God by not *speaking* to the rock in God's name, and therefore by not giving God the praise for the provision. The consequences for Moses were far reaching. By disobeying God he (and Aaron) would not enter the Promised Land. He would see it from afar, but others, namely Joshua and Caleb, would lead the people into the Land.

Moses' bitter experience at Meribah is recalled time and again in Israel's history. It comes, for example, in Psalm 95:8-9, Isaiah 48:21 and 1 Corinthians 10:4. In each case the writer reflects how Moses had become angry with the way in which the people refused to acknowledge God's ways, being only familiar with his works. Knowing God's *works* is not enough to elucidate a living faith; there needs to be a knowledge of God's *ways*, how he does things, as well as what he does.

Water flowing from a rock appears to be magical at first sight. However, it is not an unknown phenomenon in the deserts of the Middle East. Water from sparse and rare rainfall can seep through the ground surface to the soil and rock underneath. In certain cases it can fill cavities in a rock stratum, waiting to seep out through a break in the rock. Striking the rock can generate a fissure that releases the stored water. Moses may well have struck such rocks on occasions during his 40 years in the desert looking after sheep for his father-in-law. But whether due to Moses' skill or God's grace, he had been commanded by God to speak to the rock, not to strike it. That way, God would be seen to release the water. But Moses had done it himself and not given God the glory. He was to live—and die—with the consequences.

Following this experience, God is referred to as the Rock. He is the foundation on which all can stand. He is also the provider. Similarly, Christ is referred to as the Rock. The Apostle Paul in 1 Corinthians 10:1-4 says that 'Our fathers were all under the cloud and that they all passed through the sea. They were all baptized into Moses in the cloud

Water from the rock

and the sea. They all ate the same spiritual food and drank the same spiritual drink for they drank from the spiritual rock that accompanied them, and that rock was Christ—yet God was not pleased with most of them—their bodies were scattered over the desert.' The Israelites shared common experiences. They were called to love and care for one another in the desert, but they still had the attitude of slaves and were focused on their own individual needs rather than the needs of each other. They also failed to recognize God's ways in dealing with them. He continually provided them with all they needed to survive and even to flourish in the desert. The angel of God went before them on their travels: their clothes did not wear out, nor did the shoes on their feet; see Deuteronomy 29:5. Unperceived, Christ was with them, the one who was their resource, and who would one day be hung on a tree for them with his side pierced by a Roman spear. He would be the Rock from whom blood and water would stream out to resource all who turn to him; 1 Corinthians 10:4. God wanted his people whom he had rescued from Egypt to reach the Promised Land, but they failed to understand God's ways. Therefore that generation paid the price and died before they reached the Land.

Resource questions:

What lessons can we learn from the experience of the Israelites in the wilderness about not grumbling? What positive things can we do to avoid grumbling?

What is your experience of Christ the Rock?

What are the differences between knowing God's works and his ways?

11

Water and ritual washing

'Jesus . . . . got up from the meal, took off his outer clothing, and wrapped a towel round his waist. After that, he poured water into a basin and began to wash his disciples' feet, drying them with the towel that was wrapped round him.'

From John 13:3-5

Read Numbers 19

In the developed world, we are used to washing our hands with soap and water before meals or after going to the toilet, without giving such an action a second thought. The ability of the soap to remove traces of harmful substances from our hands, and of the water to dissolve and convey them away, is critical. The practice of hand washing is good hygiene and part of our culture, having been ingrained from childhood. In temperate climates, there is usually enough water for people never to have to question its importance in maintaining physical health. Yet in developing countries, where water can be in short supply, more thought and care has to be given to maintaining personal hygiene.

Washing with water is also important in religious practice. This not only promotes the personal hygiene of the worshippers, but it also symbolizes being cleansed and purified when preparing to serve God. In this latter sense, water has a spiritual as well as a physical value. An example of these dual values can be seen in the ritual that Aaron and his sons had to undergo when they were consecrated as priests to serve the LORD. First they had to wash themselves before their consecration (Exodus 29:4). Then, when they progressed from the tent of meeting to the altar, they had to wash their hands and feet with water from a bronze basin (Exodus 30:17; Leviticus 8:21; Numbers 19). Next, the animals for sacrifice were washed with water before being burnt (Leviticus 1:9). When they had completed their tasks, the priests were regarded as ritually unclean, and they again had to wash themselves and their clothes, after which they only recovered their 'clean' status in the evening. The focus was on cleansing the skin or clothing from any impurities or anything that would tend to invalidate the service of the priests or the effectiveness of the sacrifice.

The 'water of cleansing' used in the rituals was made from water mixed with the ashes of a red heifer. The heifer had to be without blemish. It was taken outside the Israelite camp to be killed, then it was burnt in its entirety; Numbers 19 records the details of what was required. The ashes of the heifer were kept outside the camp in a jar. They were mixed with water when needed to provide the 'water of cleansing'. This was regarded importantly as a means of purifying things and people from sin, especially those that had been in contact with, a human bone or grave, or a dead body, whether the body was of someone who had been killed or died a natural death. The person to be cleansed had to be sprinkled with the water of cleansing on the third day after the event and again on the seventh day. If someone died in a tent, then the whole of the tent, its furnishings and the people who had been in the tent, were similarly unclean and had to undergo the same ritual. The ritual was carried out by a man who was initially ceremonially clean. He was required to put a bunch of hyssop into the 'water of cleansing' and then sprinkle the unclean persons and objects. Even touching the water of cleansing made the man unclean, and he then had to wash his clothes and remained unclean till the evening. According to the Old Testament, the water by itself did not cleanse anything; it was a symbol of the cleansing that God brings about. The unclean person who did not carry out these tasks to the letter was cut off from the community. The general principle about uncleanness was that anything or anybody who touched or was touched by something or someone unclean automatically became unclean. Those who were ritually unclean were excluded temporarily from the community and had to go 'outside the camp'.

Water was also used symbolically to restore people who had recovered from skin disease or infections back to society. Water was not used in these cases to cleanse skin but to symbolize purification. When Jesus healed someone with leprosy, he told then to go and report to the priest so that they could undergo the necessary purification rites (with water), because that was the way in which they would be accepted back into the community.

Water was used to wash blood from clothing (Leviticus 6:27), to receive and dissolve any blood spilled when killing birds for sacrifice (Leviticus 14:5), to cleanse articles from mildew (Leviticus 14:35-57) and to deal with any form of specified uncleanness such as the discharge from a woman during her period (Leviticus 15). Although these were religious rituals, there was obviously a strong element of hygiene involved, and the need to protect the health of individuals and the community.

Ritual washing was also required after men and women had particular bodily discharges. After such a bodily discharge, anything and anyone the person touched became unclean, whether beds, seats or clothes. A couple who had sex had to bathe in water and be ritually unclean until the evening. If the discharge of semen was abnormal, or the couple had sex during the woman's menstrual period, then the situation was more serious; both the man and the woman had to bathe as for the normal discharge, but they had to wait seven days before they became clean, after which they had to wash their clothes and bathe with water. On the eighth day, a man or a woman who had an abnormal flow of blood had to give either two doves or two young pigeons to the priest to sacrifice on their behalf: one for a sin offering and the other for a burnt offering. (A woman who had her normal menstrual period did not have to carry out the sacrifice). These laws were designed to introduce restraint: sex was not to be the main focus of the relationship between a man and a woman. Certainly there was nothing evil about the discharges concerned, but no-one who was unclean could come before God in the tabernacle or temple.

The use of water for cleansing, whether as a ritual or daily habit, is prominent in most religions; it can be used as a purifier, or have spiritual and healing powers. Hinduism regards rivers as gods; the main one being the River Ganges, Mother Ganga[69]. Islam requires believers to wash certain parts of the body before performing prayers five times daily. Baptism in or with water is an essential initiatory rite for Christianity; and Judaism and Sikhism, among others, also make use of water in this way. In Judaism and Islam, a dead person is ritually washed in pure water before burial. Islam regards life itself to be integrally dependent on water: 'We made from water every living thing.' (Koran S. 21:30 Yusuf Ali[70])

Sometimes water is specially prepared for religious purposes. For example, some Christians sprinkle themselves with water that has been specifically blessed, and therefore regarded as being 'holy'. The waters of particular springs have been attributed with healing properties. Many sick people travel to Lourdes in France, where a young girl, Saint Bernadette Soubirous, received visions of the Virgin Mary in a cave; she was encouraged to dig down to discover a spring, the water of which would bring healing to those who drank the water or washed in it[71]. As mentioned above, many religions regard certain bodies of water, and even rivers, as being sacred or auspicious.

In New Testament Christianity, it is not surprising that sin is connected with maladies such as skin diseases and leprosy. In particular, lepers have to be cleansed; and if they become clear of the disease they have to undergo a ritual purification with water before they can be accepted back into the community. It follows that leprosy is often regarded as a metaphor for sin, with the subsequent need for cleansing of the sinner[72]. It is again no surprise that baptism with water is regarded as a symbol of cleansing from sin at the beginning of the Christian life (Psalm 147:3; Isaiah 1:5-6; Jeremiah 8:2; 30:12; Mark 2:17). Cleansing became one aspect of Christ's atonement and forgiveness (James 4:8; Hebrews 10:21-22; 1 John 1:7, 9). Moreover, the removal of the infected person from the inside to the outside of the camp is the parallel of the banishment of Adam and Eve from the Garden of Eden once they had sinned. In the new heaven and the new earth, believers are urged to wash their robes, when they have the right to eat fruit from the tree of life (Revelation 22:14).

References:

Lourds: http://www.catholicassociation.co.uk/lourdes/introlourd.shtml

Resource questions:

Reflect on your daily routine. How often do you use water and for what purposes? In particular, how often do you use water for washing?

What cultural issues do we have about washing with water?

What 'hidden' uses do you make of water?

Look again at the various ways in which the Law of Moses involved using water for washing people and things. See if you can distinguish between those occasions when water is used ritualistically, and where it is (obviously) used for its physical washing away of pollutants and impurities. When it is used ritualistically presumably there is a spiritual use being made of the liquid. What is the strength of this aspect of water in our lives today?

12

River Jordan and the Dead Sea

*'You could not step twice into the same rivers, for other
waters are ever flowing on to you'*
Heraclitus of Ephesus
See http://en.wikiquote.org/wiki/Heraclitus

'The river is the calling card of its catchment'
Roland Price (Author)

Read Psalm 65:9-13

Rivers are both the arteries and veins of the Earth's land surface;
Prof Patrick Denney of UNESCO-IHE, refers to wetlands as the Earth's
kidneys![73] Rivers bring a vital supply of fresh water for irrigated
agriculture and direct human consumption. They sustain the ecology
of a region, and can transport much needed sediment and silt for
improving the fertility of flood plains for agriculture. The larger rivers
facilitate communication and navigation along their corridors. Rivers
also convey salts leached from their catchments by rainfall that has
percolated through the soil, and any waste material from agriculture
or other human activity. Rivers can become so highly polluted that
one community's wastewater can destroy the water supply of another
community downstream.

Rivers are specifically recognised in the Bible as being an
important part of God's creation. The Psalms particularly celebrate
rivers: 'There is a river whose streams make glad the city of God'
(Psalm 46.4, CW:DP); 'You visit the earth and water it; you make it very
plenteous. The river of God is full of water.' (Psalm 65:9, CW:DP); 'Let
the rivers clap their hands, and let the hills ring out together before
the LORD, for he comes to judge the earth.' (Psalm 98:9, CW:DP)
Besides creating the four rivers which flowed out from the Garden of
Eden to water the earth (Genesis 2:10), God 'cleaved the earth with
rivers' (Habakkuk 3:9). In restoring Israel after the Exile, God states

that He is 'making rivers in the desert' (Isaiah 43:19-20). He will open rivers on the bare heights (Isaiah 41:18), and will lead the people of Israel by streams of water (Jeremiah 31:9). The river Euphrates appears several times as a significant boundary for Israel's ambitions; see Genesis 15:18, Deuteronomy 1:7 and Joshua 1:4.

The Promised Land was blessed with 'early and latter' rains. Such rain softened the land and filled its furrows at the right time for corn-based agriculture, the planting and fruitfulness of vineyards and the support of flocks of sheep and goats. But the Land also had an important river of its own: the River Jordan, which formed its eastern boundary. The Jordan is an unusual river because it flows along a rift valley, on average about 10 km wide, formed by movements of the Arabian tectonic plate in the earth's crust. The valley actually extends from Turkey in the north to East Africa in the south. The lowest point of the valley is by the shore of the Dead Sea at the downstream end of the Jordan River. This point has the unique property of being also the lowest part of the Earth's surface (on dry land), at about 422 m below sea level[74]. The land to the west and the east of the rift valley along the River Jordan rises to about 1000 m above sea level. Rainfall on the western slopes infiltrates the ground to feed aquifers that drain eastwards towards the Jordan, which runs from north to south along the rift valley.

The source of the Jordan River is 75 km north of what used to be a swampy lake called Lake Hula, which was marginally above sea level. Lake Hula was drained in the 1950s and the land is now used for agriculture. Here the rainfall is about 550 mm per year. From the site of this lake, the river flows steeply for 25 km to the Sea of Galilee, whose surface is 209 m below sea level. This inland lake is about 21 km by 13 km, has an area of approximately 166 km² (about 8% of the area of the Ijsselmeer in The Netherlands), and has a maximum depth of about 43 m[75]. The Sea of Galilee (or Lake Tiberias or Lake Kinneseret) is the lowest fresh water lake on earth.

The Jordan meanders from the Sea of Galilee for about 217 km to the Dead Sea, though the actual distance straight down the valley is only about 105km. The Yarmouk River, which is the main tributary of the Jordan, joins from the east just below the Sea of Galilee, and there are some other minor tributaries along the reach. There follows a portion of the river valley some 24 km across known as the Ghawr

(Ghor), which is intersected by ravines and wadis looking something like a lunar landscape. The concluding reach of the Jordan is called the Zur. The Jordan receives most of its water from rainfall running off the plateaus to the east and west of the rift valley. Besides the identified tributaries, there are a number of springs on the western bank coming from the sizeable aquifers. Indeed, there are thermal springs at Hammat Grader on the west side of the Sea of Galilee, which contribute significant amounts of salts dissolved in their water[76]. The cumulative flows to the Jordan during the winter period (January to March) meant that the Zur used to flood frequently, had a rich ecology and had considerable agricultural potential. So in Joshua's day, the Jordan River regularly overflowed during harvest (Joshua 3:15). Job found it worth commenting that 'the raging of the Jordan does not upset the hippopotamus' (Job 40:23).

The Dead Sea is about 67 km by 18 km, with a surface area of 1400 km^2 (about 70% of the area of the Ijsselmeer in the Netherlands) and is the final destination of the River Jordan. A small delta distributes the river water into the Sea. The water surface is about 422 m below sea level, and the maximum depth is about 385 m[73]. Because there is no outlet, and evaporation from the Sea is high, the water level remains reasonably constant. As a consequence, the salinity is very high at 33.7%, which is sufficient for a person to float on the surface and read a newspaper without it getting wet! The Dead Sea is a valuable source of potash, and it also has a number of naturally occurring asphalt pits, which result from volcanic activity in the area.

The Jordan River has always been an attractive area for human settlement. Jericho, the first city captured by the invading Israelites, was situated to the north of the Dead Sea. It is very likely that the cities of Sodom and Gomorrah, destroyed in the time of Abraham and Lot (Genesis 18), were located in the region, as were Admah, Zeboim and Zoar; see Deuteronomy 29:23. Zoar escaped destruction when Abraham's nephew Lot escaped to Zoar from Sodom; see Genesis 19:21-22. In Roman times, some Essenes settled on the western shore of the Dead Sea; they have been identified with Qumran and the Dead Sea Scrolls discovered during the 20th century in nearby caves[77].

Jesus spent much of his ministry in the region of the Sea of Galilee, though he was familiar with the Jordan down to the Dead Sea. Indeed, he was baptised by John in the Jordan River. Precisely where he was baptized is not known, though sites favoured by scholars include the east side of the Jordan near a place called Bethany (John 1:28) and the west bank of the river just south of the famous Allenby Bridge, constructed in 1918 by the British General Edmund Allenby and destroyed and rebuilt in 1946 and 1968[78].

Like the Nile, the Jordan also experienced floods, but on a much smaller scale. In recent years, floods in rivers around the world have come more to our attention. There have been huge floods in the Yellow River and the Yangtze in China, the Mississippi in North America, the Indus in Pakistan, the Ganges in India, and the Mekong in Vietnam, to name but a few. 2012 was the wettest year on record in the United Kingdom, with many people suffering from floods in different parts of the country. The ability of engineers to predict, limit and manage flooding is important for many communities. Today, computer models of unsteady flow in a river can realistically describe and predict how floods will develop as they move along the river. These models are based on the application of Newton's laws of physics to the flow of water at a point, and the resulting numerical equations are deduced for flow in three, two or one spatial dimensions; time is yet another dimension. The computer models confirm that the flood peak moves downstream and, provided there are no tributaries and any flow from groundwater along the reach is negligible, the peak discharge decreases. In this way, water engineers can design appropriate flood risk management strategies or develop forecasting algorithms to implement flood warning systems.

Models need data if they are to be applied successfully. But data is valuable in its own right. For example, we may be interested in the risk of flooding at a certain point in a river. Such a risk needs to be determined in terms of the statistics of the rainfall, the frequency of water level or depth exceedance, and the costs of damage due to flooding, be it goods or property directly affected, or the limitations imposed on the emergency services or people living in the flood plain. This view of data can result in the relevant databases being of key importance, while models are used to complete missing data in answer to what-if questions, or to fill in data at other geographical points where no data was collected previously, or to complete time series containing missing data. Data, information and knowledge are critical if we are to understand the behaviour of water in the natural environment. In some cases where we lack time series data, provided we know the physics of what happens and have a detailed description of the physical geometry, we can generate the missing data through models constructed using our detailed knowledge of the physics. In the last 70 years the ICT revolution has unleashed a torrent of data in all sorts of different areas of interest. The manipulation of this data stream by a vigorous and increasingly sophisticated technology threatens our sense of reality and diverts us into a myriad of virtual, digital environments. Whereas we are adaptable, we need to identify and resist the demons in the digital darkness and strive to walk in the light of God's reality whether continuous or discrete[79].

References

CW:DP: *Common Worship: Daily Prayer,* copyright © the Archbishop's Council, 2005

Resource questions:

The River Jordan is an outstanding feature of the region and of biblical history. What geographic features, if any, have played an important part in your life? What influences have they had on you? In what ways is your history tied in with your physical location?

13

Crossing the Jordan

'When I tread the verge of Jordan
bid my anxious fears subside.
Death of death and hell's destruction
land me safe on Canaan's side.'
From the hymn 'Guide me Oh thou
great redeemer' by William Williams, 1717-1791
See http://en.wikipedia.org/wiki/Cwm_Rhondda

Read Joshua 3, 4

Moses was dead. So were all the men over 40 years old who had been around when the Israelites had balked at the prospect of going into Promised Land and taking it by force; all, that is, except Joshua and Caleb. Although the Land was fruitful and prosperous, the Israelites were dismayed by the inhabitants, whom they viewed as being far too strong for them to conquer, even though they were assured that God would be with them when they went into the land to occupy it. The Jordan River formed the eastern boundary of Canaan, the heart of the Promised Land. The boundaries to the larger land promised to Abraham included the Great Sea on west (the Mediterranean), the River Euphrates to the east, Lebanon to the north and the desert to the south. Boundaries can form considerable physical or psychological obstructions, and the River Jordan in flood was not a trivial obstacle to the thousands of the Children of Israel with their families, animals and other possessions. Two and a half tribes, Reuben, Gad and the half tribe of Manassah, voted to stay on the east side of the river Jordan, where they saw much that attracted them. Moses agreed to their choice, provided they assisted their brothers in taking the Land God had promised; see Numbers 32:20-24.

Joshua was well schooled in the ways of Moses and of God. God gave him explicit instructions that the priests were to take the ark before the people and to step into the water. They were promised that when they did so the river would stop flowing and a path would open up, in the same way as had happened when their forbears had crossed the Red Sea, escaping from the Egyptian army. In other words, this new generation was to have a similar experience to their parents. They were not now being pursued by an enemy, but they had a challenge before them.

The priests did as they were told, and the flood waters stopped flowing. It was a miracle, like the crossing of the Red Sea. A physical explanation of the event could be that there had been an earthquake or landslide which temporarily blocked the river some distance to the north; the Israelites would have a limited time to cross the river before the waters built up behind the blockage to such an extent that the barrage would be broken and swept away. The waters would then return to their normal flood flow.

The people marched across the Jordan in an orderly fashion. One strong man from each tribe was ordered to pick up a stone from the middle of the river. They used these stones to build a monument to commemorate this important occasion and to be a sign to succeeding generations of the fact that the Israelites had crossed the boundary and taken possession of the Land. There was to be no turning back; they were only to look forward. The Israelites had come home. For the last 40 years they had not known what it was to remain in one place; now they would exchange their daily dependence on God as provider of water and manna for the opportunity to grow their own food as they worked the Land that God loved. They would still be dependent on God, but in a different way. The land would yield its fruit as a result of both the work the Israelites put into it, and the early and the latter rains that God would essentially provide. For the people to flourish in the Promised Land, they would have to work in partnership with God, just as Adam and Eve had been told to tend the Garden of Eden. Would the Israelites make a better job of caring for the Promised Land than Adam and Eve did of the Garden? Actually, it would not be so much their caring for the Land as their obedience to God, for Israel was to reveal the LORD to the nations. Just as Adam and Eve were to obey the LORD, so the Israelites would need to observe the laws and commandments that the LORD had given Moses. The Bible reveals that just as Adam and Eve disobeyed the LORD, so Israel consistently failed to keep the Law, and that brought about their steady decline following the golden age of King David and King Solomon. But the LORD had in mind a much longer term plan that involved his only Son, the Christ, the Messiah. His death, immediately outside Jerusalem, and his subsequent resurrection on the third day would open up a new and more effective way for all people to be reconciled with the LORD.

Jesus' death and resurrection are viewed in the New Testament as a crossing from death to life, whether symbolically in baptism, or in actual physical death. The boundary crossed in baptism is not to be traversed in reverse. In other words, the Christian once baptized cannot be unbaptized; he or she becomes a follower of Christ forever. Similarly, as the last verse of the great hymn 'Guide me O Thou great Redeemer' states, the Jordan has long been regarded as a symbol of the boundary between earth and heaven, in other words, of death, which again is irreversible.

Many peoples in the Church world-wide, especially in Africa, South America and Asia, are grasping the opportunity to enter their Christian inheritance, often in the context of miracles accompanied by a vigorous faith, as well as by persecution for some. It is only in Europe, and to a lesser extent North America, where churches are in decline because members appear to have lost courage and the power of the Holy Spirit in their lives, that Christians are fearful in the advance of a militant secularism. We need urgently to recover our sense of a God who works miracles, and who calls us to follow him.

Resource questions:

Are we the advance guard of our fellow Christians elsewhere, or do we need to regain our courage and boldly go into our Promised Land afresh?

How would you use the account of crossing the River Jordan to encourage somebody facing death?

14

A cup of water

> 'A traveller on foot in the desert despaired of ever
> finding water. . . . Then he heard something, so faint that
> only the sharpest ear and the deepest silence would lead
> to its detection: the sound of running water.'
> In 'The heart of the enlightened' by Anthony De Mello

Read 2 Samuel 23.8-17

A cup of water after long, hard cycle ride in the warm, summer sun, following winding cycle paths through the countryside[80], is more than just a refreshing drink: it is symptomatic of our basic need for the liquid that makes up the greater proportion of our bodies. We sympathize with the Psalmist (Psalm 42:1) in linking his desire for God with the deer panting for the water brooks where it may slake its thirst. A drink of water, or 'Adam's ale', at the right time can be better than the sweetest, smoothest nectar. It can also be a life saver, and in giving a cup of water to someone in need there can be a profound expression of compassion and community.

Jesus leaves us in no doubt that we are not only to love our neighbour as ourselves, but to love those with whom we are in deep conflict, namely our enemies. 'If (your enemy) is thirsty give him water to drink' (Proverbs 25.21). And you will be blessed in giving to one of the least of Jesus' followers a cup of water because he is a follower, for in doing so, you will receive a reward; see Matthew 10:42. Job was accused unjustly by Eliphaz of not giving water to the weary (Job 22.7). Fresh water is a basic human good, and a basic human right, with repercussions far beyond the normal interaction between human beings.

What is more, God is involved. He provides for us in every situation when we believe in him and call upon his name. He is concerned to give us the major things, such as eternal life and the little things, such as our daily food. But there is really no large or small: everything he gives us has its unique value and blessing. This principle is highlighted in Judges 15:19 where Samson, God's strong man, had lost strength and needed water. God opened up a hollow place in Lehi, and water came out of it. Samson drank from the spring called En Hakkore, and his strength returned to him. In asking God for a drink, Samson knew that the Spirit who empowered his strength for battle against the Philistines was the same source he needed to sustain his physical life.

There is another heartwarming story about a cup of water in 2 Samuel 23.8-17. King David was thirsty, and expressed a desire for water from the well near the gate of Bethlehem, his home town, which was at that time occupied by the Philistines. Three men overheard his wish. These were no ordinary men: they were the three bravest, strongest and most valiant of David's soldiers: Josheth Basshebeth, Eleazer and Shammah, the *creme de la creme* of David's fighting force. They were the three mighty men. Their love and respect for David knew no bounds. At great personal risk, they broke through the Philistine lines, reached the well and drew water from it. Whether they were seen or noticed by the Philistines we are not told, but having secured some water in a container they made their way back to camp and brought it to David. When David knew where it had come from and how it had been obtained, he refused to drink it. Instead, he poured the water out on the ground before the LORD. To have drunk it would have been, for him, as if to drink the lives of his men who had secured it at such risk to their own lives. David respected the gift of his men and their love for him, but he turned his personal desire into a sacrament of thanksgiving for the bravery and courage of the three mighty men.

Perhaps the most poignant request for a drink of water in the Bible is Jesus' request to the Samaritan woman (John 4.7). After his baptism, Jesus and his disciples hung around the Jordan where his disciples began to baptize people who came to them. This provoked serious attention from the Pharisees. When Jesus realized he was under threat, he set off with his disciples to Galilee. They reached Jacob's well in Sychar at midday, and the disciples went into the village nearby to get some food, leaving Jesus to rest at the well. It was hot, and Jesus was tired and thirsty. Being midday, there would not normally be anybody drawing water from the well at that time, so when the disciples left Jesus he was probably alone; but not for long. A woman came out of the village to draw water.

She had not come earlier because she was something of an outcast in the village: it was more comfortable for her to come alone in the heat of the day. But on this occasion, there was a man resting at the well, and the woman approached cautiously. Without a word she began to draw water, averting her eyes from the stranger, who was obviously one of those pompous Jews. And to her great surprise and consternation he spoke to her. Instead of criticizing or rebuking her, he put himself in her debt by asking for a drink of water. She had the capability of giving him such a drink from the well, but the Samaritan woman was quick to respond: 'You are a Jew and I am a Samaritan woman. How can you ask me for a drink?' The social and political divide between Jews and Samaritans was so strong, it denied her even the opportunity to share water. But she was intrigued. Her curiosity led to her to listen to what this enigmatic Jew had to say to her. She found that he knew far more about her than she believed possible, and this opened up the way for her to embrace new life; see John 4:39-42.

Today in many countries, we are spoilt for choice when we need a drink. All sorts of bottled beverages are available, whether from big multi-nationals or local suppliers, who offer carbonated or still water with a wide range of flavours. Indeed, plain bottled water has become a huge industry world-wide with a value in excess of $100 billion. Global sales of bottled water exceed 200 billion litres annually[81]. In many cities of the developing world, municipal water supply is intermittent and fails to meet adequate water quality standards, so demand for bottled water is high. But demand is also high in developed countries where, according to the World Health Organization, bottled water can have a greater bacterial count than municipal water[82]. For some reason, in a restaurant, we prefer to have bottled water on the table rather than a jug of water from the tap. It is somewhat ironic that the Food and Drug Administration in the United States has lower standards for bottled water which it regulates, than for tap water regulated by the Environmental Protection Agency[83]. Do we, in developed countries, consider it therefore more refined to offer somebody a drink of water from a bottle rather than filling the cup from the tap? And should we be promoting an industry in which only about 20% of the plastic water bottles (made usually from polyethylene terephthalate) is actually recycled?[84]

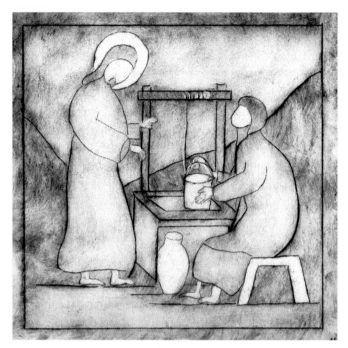

Water from the well

Of course, it is different in developing countries where it is probably advisable not to drink municipal water straight from the tap. Here the water should preferably be boiled to kill off bacteria. Another way to make the water safe for drinking is to put it in a plastic water bottle in the sun for 24 hours; in many countries there is sufficient ultraviolet light in the sun's radiation for it to make the water safe. This approach to purifying water has been queried by scientists however, who warn against that carcinogenic substances from the plastic may be dissolved into the water, particularly if the plastic bottle is reused many times. They also advise that the water should be consumed soon after removing it from the sun.

Ultimately, it is of God's nature to provide for his people, water and food as well as safety, security and sustenance. Just as he provides the water brooks for the wild animals (Psalm 50.1) and still waters for those dependent on him (Psalm 23), so he provides water and food resources for his servants, such as the children of Israel at Meribah (Exodus 17.7) and Elijah after his confrontation with the prophets of Baal (1 Kings 19.6). The clarion call of Isaiah the prophet in exile was 'Come to the waters, come buy without money, come buy and eat' (Isaiah 55.1). Isaiah's vision is fulfilled in the new heaven and earth, where the call is again to come and be satisfied with the spring of living water; see Isaiah 55:1, 2.

Resource questions:

Are you ready to 'come to the waters and drink' what Jesus offers?

In what ways have you served others this last week? Have you allowed others to serve you?

How can we show more compassion towards those in need?

15

Water and hospitality

'Brother, sister, let me serve you,
Let me be as Christ to you;
Pray that I may have the grace to
Let you be my servant too.'
        From 'The Servant Song' by Richard Gillard,
        in 'Scripture In Song', Maranatha! Music , 1977

Read Gen 18:1-15

Eating out in The Netherlands, as in most cultures, is a social occasion. When I go with family or friends to a Dutch restaurant the waiter or waitress will first ask us what we want to drink. After bringing our order, we are left to enjoy conversation as we sip our drinks and decide on the menu. It may be half an hour before anyone returns to take our order for food. Dutch hospitality in restaurants is not hurried; so much so, that with my English desire to keep things moving along, I sometimes wonder whether we are being ignored by the waiters. But I am aware that the Dutch love to chat when they meet for dinner.

You would certainly not be ignored by the hospitality shown in Middle Eastern culture. And water forms a very important part of that hospitality. In Jesus' time and before, people wore sandals when walking along dusty roads. So it was inevitable that their perspiring feet would attract dust and dirt, which, if not washed off, would be offensive. It was the obligation of the person whose house they were visiting, to provide washing facilities for his guests, and in particular, water for their feet.

In Genesis 18:1-15 we read that Abraham offered water to three men who visited him when was sitting at the entrance to his tent in the middle of the day by the great trees at Mamre. Abraham became aware of some men standing nearby, and his first response was to make them welcome. It appears he knew immediately that the three men were an epiphany (that is, an appearing) of God. Abraham bowed down to the ground before the men. He called his servants to bring water and food for the men to wash and eat. Then the LORD gave Abraham and Sarah the all-important news that the childless Sarah, who was now well past the age of bearing children, would have a son in about a year's time. Isaac was the son of the God's promise to Abraham that from his descendents there would be one through whom God would bring blessing to all nations.

There are two important occasions in the Gospels where Jesus was involved in foot washing. The first occasion was when he received the attentions of a 'woman of the street'; see Luke 7:36-50. In fact, we know her name as Mary. She came to find Jesus while he was being entertained to dinner by a Pharisee called Simon. Jesus' host had singularly failed to be hospitable to his guest. During the meal, as Jesus was reclining on a couch at the table, Mary freely entered the house, (a custom of the day) and knelt behind Jesus. Without saying anything, she began to let her tears fall on his feet, then wiped his feet with her hair and began to kiss them. Then she took an alabaster box, which contained expensive ointment, poured out the ointment on his feet and smoothed it into his skin.

Meanwhile, the host, Simon, watched with amazement and not a little disdain. How could this rabbi allow a woman, and such a woman, to touch him? Didn't he know who she was? Certainly, it seemed to Simon that Jesus was unaware of the people around him. Or was he? Jesus made it clear that he knew what Simon was thinking. He told him that he would not defend the woman's private life, but he contrasted what she was doing with what Simon had failed to do. Simon, as host, had not washed his feet, had not kissed him when he had arrived, and had not anointed his head with oil. But this woman had not ceased to wash his feet with her tears, kiss them and anoint them with oil. Who had shown Jesus true hospitality?

Why did the woman do what she did? Her tears were probably a strong indication that she was deeply sorry for her past life. By focusing on Jesus' feet, she had found a way of expressing her repentance. Jesus assured her that her sins, which he said were many, were forgiven. Her faith enabled her to move on to live as a disciple of Jesus. Mark saw this event as the trigger that persuaded Judas to betray his Master (Mark 14:10-11). Judas failed to appreciate the lavish attention that Mary gave to Jesus, and begrudged the value of her gift.

The second important occasion in the Gospels was when Jesus washed his disciples' feet; see John 13:2-17. The twelve disciples had gathered with Jesus to celebrate the Passover meal in an upper room. Two disciples had been sent ahead by Jesus to make the necessary preparations for the Passover. All was ready. But there was an uneasy atmosphere in the room because the traditional service of washing feet had not taken place, and who was there to do it anyway? Normally foot washing was the task of the lowest servant or slave. But who was that person among them? There was no acknowledgement that any of the twelve was willing to take responsibility for this most menial of tasks. Indeed, they often seemed to spend their time arguing about who was the greatest among them; who would have positions of authority when Jesus' kingdom came into being? So their feet had not been washed—they were dirty and smelly. And then, to the amazement of all present, Jesus took a basin of water and a towel, laid aside his garments, wrapped the towel around him like a servant, and began to wash the disciples' feet. They were too embarrassed to say anything, but let him pour water on their feet and wash off the dirt and sweat with his hands. That is, until he came to Peter, who had been thinking agitatedly about what he would do when Jesus reached him. He burst into his complaint: he would never let Jesus wash his feet; it was not right that his master should do such a thing to his disciple. But Jesus was not going to be put off. He rejoined by simply telling Peter that he would have no part in Jesus if he did not let him wash his feet. Peter immediately realised what Jesus was saying. His love for his master went way beyond his embarrassment at what he was then experiencing, and he blurted out that he would willingly be washed all over by Jesus if that is what he wanted. Peter went from one extreme to the other. But Jesus, by his action, had a very simple, direct message for his disciples, including Peter. He wanted them to realise that if he was their master and he washed their feet, they should be prepared to wash one another's feet. The uneasy tension in the room on their arrival could have been averted by one of the disciples performing the act of courtesy for the others, and any one of them could have done it, but their failure gave Jesus the opportunity to teach them a very important lesson.

Later, Jesus went so much further in serving his disciples when he laid down his life for them—and us. He calls on us to give our

Jesus washing his disciples' feet

lives in extravagant acts of selfless service. Remember, Jesus must have washed Judas' feet, knowing that Judas was about to betray him. Jesus did not give the other disciples any indication that the relationship between him and Judas was strained. Such is the love Jesus has for each of us.

Resource questions:

How prepared are we to serve our brothers and our sisters?

What can we do to serve the brother or sister we find it difficult to get on with?

16

Water and urban society

'Judah mourns, her cities languish; they wail for the land, and a cry goes up from Jerusalem. The nobles send their servants for water; they go to the cisterns, but find no water. They return with their jars unfilled; dismayed and despairing, they cover their heads'
                                                    -Jeremiah 14:2-6

'The Romans realized, as have every civilized people since, that living in cities is impossible if the water supply is not reliably clean and fresh."
                            -Frank and Francis Chapelle,' The Hidden Sea:
                                Ground Water, Springs and Wells", 1997

'The coolness of water quenches thirst, its motion cleanses, its vigour gives or restores health. In a vaginal douche, it (water) is privy to the 'darkest secrets' and, so it was believed, acted as a means of birth control; lukewarm and perfumed it offers pleasure that used to be reserved for dandies; in a village fountain it takes part in the life of the community and directs the ballet of the women who meet around it; its washbowls and basins, in bidets and water closets it permeates the rites of cleanliness and hygiene, in the sewers it cuts through the entrails of the city; in a water tower, it changes the skyline.'
                            Coubert, J-P, The Conquest of Water, Princeton
                            University Press (translated by Andrew Wilson)

Read 2 Chronicles 32:27-33

By 2008, 50% of the world's population lived in cities, and the percentage is slowly increasing[85]. Jerusalem, the city of David, had between 30,000 to 50,000 people living there permanently during Jesus' time. The population swelled to between 80,000 and 120,000 during major religious festivals. The provision of fresh water in the city was therefore very important. Water was provided in various ways. Rainfall was harvested by collecting the runoff from roofs and storing the water in cisterns; these were lined underground tanks, often carved into the rock. There were also wells reaching down to water in the aquifer below the city. These wells provided some guarantee that the inhabitants would have access to water during a siege.

A dramatic picture of the situation in Jerusalem during a drought is given by Jeremiah (Jeremiah. 14:3). Tunnels that rerouted fresh water streams into the city were an important means of providing additional water to the population. King Hezekiah had realized the importance of these tunnels (2 Kings 20:20; see also 2 Chronicles 32:4, 10 and 2 Kings 18:17). When he heard that Sennacharib, the King of Assyria, was coming to capture Jerusalem, Hezekiah blocked off the water from the springs outside the city, particularly the Gihon spring, and channeled the stream that flowed down the west side of the city through a 1750 ft (0.533 kms) long tunnel under the city wall into the city centre; see 2 Chronicles 32:3-5, 30. The tunnel fed the Pool of Siloam which features in John 9:11. The pool and its tunnel were no mean feat of engineering. The technology necessary to construct them was very advanced for the end of the eighth century BC. The Siloam inscription, which had been carved in the rock, was discovered in 1880[86]. The carving of the inscription marked the end of the project. Hezekiah certainly did not make life easy for the intending invaders. He was also responsible for an aqueduct from the Upper Pool on the road to the Washerman's Field; see Isaiah 7:3.

A significant water supply facility was constructed during the Maccabean era at the time of the reconstruction of the temple at Jerusalem[87]. This became known as Solomon's Pools, about 5 km southwest of Bethlehem. These consisted originally of three open cisterns fed from an underground spring, each pool being over 100 m long, 65 m wide and 10 m deep. They were so designed that the bottom pool filled first, and the others in succession. There is a 6 m drop between the pools, which helps to aerate the water, that is, to increase the dissolved oxygen. Later, Pontius Pilate built 39 km of aqueduct to carry water from the collection pools at Arrub. Herod the Great created the underground tunnel feeding the upper pool. An aqueduct system conveyed water from the pools to Bethlehem and to Jerusalem, where the aqueduct terminated under the Temple Mount.

The present-day water system in Jerusalem is sophisticated[88]. The pools are fed by four separate springs; the most prominent is called 'the sealed fountain' at the head of the Wadi Urtas, about 200 m to the north-west of the upper pool. The spring water is transferred to the upper pool by a large subterranean passage. There used to be five aqueducts distributing water from Solomon's Pools. These had a total length of nearly 60 km. One aqueduct from the lower pool took water through Bethlehem and across the valley of Gihon, then along the west slope of the Tyropoeon valley, until it discharged to the great cisterns underneath the temple hill in Jerusalem. Nowadays, the water only reaches Bethlehem, because the remaining sections of the aqueduct have been destroyed. Two of the aqueducts brought additional water from the south, and another aqueduct carried water from the upper pool east to Herod the Great's desert fortress, the Herodium, where he had constructed a large recreational pool, lined with columns. Finally, two aqueducts brought water from Solomon's Pools to Jerusalem.

Today, Jerusalem has a population of more than 650,000, and treated fresh water is distributed to every house and building through a network of pipes operating under pressure, as is done in any typical modern city. The important feature of the distribution network is that it operates under pressure: if the pressure fails for any reason, such as a pump failure or a pipe burst, then polluted groundwater may enter the pipe network and affect the treated water in the system, thereby making the water unsafe for drinking. This is often the case in developing countries where electricity power failures lead to intermittent water supplies. What is more, because the pipes operate under pressure, any weakness in the joints between the pipes, or cracks in the pipes themselves, is exploited as water seeps through them and is lost to the ground around the pipes. This is a complete loss to the system, and although some loss is inevitable in a large network of pipes, leakage of more than 30% can be unacceptable. This is especially true in arid climates where there is a shortage of fresh water.

The supply of fresh water through networks of pipes underground is comparatively recent. Up to the eighteenth century in Western

Europe (particularly France and Britain) water in urban areas was provided by water carriers[89]. As the underground pipes brought water directly to (rich) people's homes, so societies' perception of water changed from being a gift of God or nature and a privilege reserved for the nobility to being the property of everyone; see Jean-Pierre Coubert (1989). Gradually, almost subversively, water became a vital resource for the industrial revolution of the eighteenth and nineteenth centuries in England and then Europe. With the inevitable overcrowding of workers in the new industries came the demand for better personal hygiene at all levels of society. Water was not only supplied to industry, but to hospitals and schools, and ultimately, the home. Private and municipal companies competed to bring water to everyone through their pipe networks. 'Water thus conquered man in a triumph linked to increasing industrialisation and an economy that devoured water.'[90] The sacredness and spiritual aspects of water gave way to its secularisation by scientists, engineers and physicians, with cleanliness and hygiene as the primary values and drivers. Professional, scientific and technical knowledge came to stand against the search for magical powers as symbolised by the water diviner's dowsing rod, and the sacredness of water pertaining to life and death in the prevailing Christian faith. Too often the attitudes that these three ages (science, magic and spirit) have had to water have been viewed in historic sequence and in opposition to each other. We would do well to encourage reflection on the different benefits that each age brings to our human society and its intimate dependence on water.

The supply of fresh water for domestic and industrial use leads to the production of wastewater which contains different materials and liquids introduced during the water's use. The wastewater therefore includes both particulate and dissolved pollutants. Something has to be done with the wastewater. The sheer volume of wastewater can pose problems when it comes to disposal. Additionally, the pollutants in the wastewater may be a threat to human health[91]. Ideally, as much of the pollutants as possible should be removed from the wastewater before the effluent is discharged back into the natural environment. Although a limited amount of wastewater and waste material can be dumped into pits, allowing the water to infiltrate into the ground, such a facility is very limited, and most of the wastewater has to find its way back to the natural surface water streams. In a modern city, wastewater is collected directly from houses and buildings through another, separate pipe network, and conveyed under gravity either to a wastewater treatment plant where the various pollutants are removed using a number of different physical, chemical and biological processes, or discharged directly to a stream or river[92]. Many of the chemical and biological treatment processes have only been widely available in the last 150 years[93]. These processes are identified as being in three categories: 'primary' treatment, which involves the removal of solids through sedimentation, 'secondary' treatment, which aims at removing nutrients and is normally based on biological activity in what is called an 'activated sludge' process, and 'tertiary' treatment as a polishing stage for the removal of remaining hazardous chemicals and which involves more advanced chemical processes. Before these processes were available, wastewater was disposed of into the environment in the easiest way possible. Not unnaturally, this often meant that wastewater flowed down the centre of the street on its way to the nearest stream. Consequently, the stream became heavily polluted, which made life difficult for communities downstream who wanted fresh water from it. Some enterprising communities that generated a small, but none-the-less significant amount of wastewater, diverted it through reed beds that were specifically designed and constructed in shallow ponds. The reeds (or other aquatic plants) could extract pollutants from the wastewater, such as nitrates, phosphates and heavy metals. Plants in the papyrus family are now widely used for this purpose.

Excess surface water in urban areas due to significant rainfall can also cause problems of disposal. In this case, the runoff from the rainfall can be so large that it is unable to infiltrate into the ground because of the paved streets, roads and buildings covering the land surface. This water collects in depressions on the ground surface and can cause flooding. One solution to this problem is to deliberately create above-ground flow paths, so that flood water flows to the nearest stream as quickly and safely as possible. A modern city does this by either using larger pipes in the wastewater network to convey both the continuous inflow of wastewater and the intermittent stormwater flows, or building a separate pipe and channel network for the stormwater. Jerusalem is hilly, and intense rainfall is comparatively rare, so excess rainfall-runoff can be accommodated by the underground combined pipe network and above ground flow paths, often along streets. Of course, no matter how big the pipes are in the drainage system, there is always the chance that a storm with rainfall of sufficient intensity and duration will cause flooding in the urban area served by the system. The skill of the drainage designer is in recommending the correct size of pipes, so that significant flooding occurs only with a frequency acceptable to the community. This means that there has to be a tradeoff between the cost of constructing, maintaining and operating the drainage system against the damage cost (to property, human health and stress) incurred by flooding. The tradeoff can be explored by simulating the above-ground flood flows[94]. Ideally, due to the significance of the attendant flood risk, the local community should be involved in making the decision on the appropriate standards for design of the drainage system. This is a matter of social justice, and brings both water professionals and people from the community together[95]. Increasingly, good design rests on the use of computer simulation modelling of the flows in the pipe networks, both to prescribe the pipe diameters and to predict flooding and the associated damage for rainfall events rarer than the design event. Such designs and simulations may include the operation of the treatment works and the impacts on the receiving waters. Today it is possible to optimize the design and operation of the whole wastewater system in order to get the best performance for a minimum cost.

For further reading:

Butler, D and Davies, J W (2011) *Urban drainage*, 3rd Edn Spon Press

Goubert J P (1989) *The Conquest of Water*. Polity Press

Price, R K and Vojinovic, Z (2011) *Urban Hydroinformatics*. IWA Press

Vojinovic Z and Abbott M B (2012) *Flood risk and social justice*. IWA Press

Resource questions:

With the continuing migration of people from rural areas to cities, growing pressure is going to be put on the collection and treatment of wastewater and stormwater (let alone the provision of potable water for domestic and industrial use). Can you think of alternative ways of treating wastewater? How can the costs of treatment be reduced?

How can we as individuals improve the management of water in our own homes?

17

Bitter waters made sweet

'Filthy water cannot be washed.'
West African Proverb

'Only one quarter of the 21 billion tons of China's annual output of household sewage is treated . . . . Treatment plants are being built, but will still handle only half of all city sewage, leaving rural waste water untreated.'
Associated Press, 'China issues dire environment report,' MSNBC.com, 06 Jun 2003

Read Exodus 15:22-26

At the beginning of their travels in the wilderness from Egypt to the Promised Land, the Israelites were faced with a need for water, but the only water they came across at a place called Marah was undrinkable. God told Moses to throw a particular piece of wood into the water, which made it sweet (Exodus 15:22-25). Elijah and Elisha also made waters drinkable (2 Kings 2:19-22). In each case they added something to the water that dealt with the contamination. What Moses, Elijah and Elisha did was the naive precursor of our modern approach to water treatment.

My wife, Thea, and I spent a year at the University of Florida in Gainesville, where I had a Visiting Assistant Professorship. On our first Sunday in Florida, we were asked if we would like to go 'tubing'. This involved floating on a large inner tube down a spring-fed stream through dense tropical undergrowth. The water was a constant 72°F (22°C) and very clear and pure. It was so clear that we could see the grains of sand on the bed, the ever dancing, sun-refracted shadows created by bright sunlight beaming through the surface waves, and small shoals of brightly coloured fish. We spent a relaxing afternoon floating beneath a speckled green canopy, trailing cans of beer in the water behind us. (We were also challenged by the thought that there were deadly cottonmouth snakes in the vicinity of the channel, but we did not see one!) This stream was a far cry from the murky brown soup of some urban rivers draining irregular channels, complete with discarded household and industrial rubbish. These rivers, far from being a recreational delight to their communities, have become repositories for anything that the communities do not want. Poor people have to resort to living on the banks of these rivers when there is nowhere else for them to go.

Pollution is one of the major scourges of natural fresh water supplies world-wide. The danger of drinking untreated water is a serious problem. Many people fall ill and die from drinking polluted water each year, and children are especially vulnerable. Yet there are a number of ways of treating water cheaply and effectively so making it fit to drink. One of the most simple and direct ways is to boil the polluted water for a few minutes. During the industrial revolution in England, the fact that water had to be boiled to make the new drink, tea, greatly reduced illness caused by drinking polluted water.

As the industrial revolution has gained pace recently in many developing countries, water managers have turned a blind eye to pollution in the aquatic environment, focusing on profit rather than sustainability. Similarly, as the human population has increased and agricultural food production has been improved through the application of fertilizers and insecticides, so more and more toxic chemicals have entered the food chain, bringing long term, unknown health hazards. The problems posed by these and other pollution sources, including domestic waste, led to the formulation and agreement of a range of Millennium Development Goals[96]; these aims to halve the number of people without access to safe drinking water and without safe sanitation during the ten year period between 2005 and 2015. The number of people without these basic rights was estimated to be almost one and two billion respectively.

There are various sources of fresh water, including groundwater, but rivers are a primary source of water for the human populations in many developing countries. Yet many rivers world-wide have been or are currently experiencing serious pollution. The River Jordan is no exception. It has suffered severely in the race to improve agricultural production in Israel and Palestine. Martin Palmer, a journalist writing in the Guardian, described how pollution forced the closure of the river Jordan to Jewish, Christian and Muslim pilgrims[97]. He spent two days talking with environmentalists in Jordan about their country's severe water problems. Both the river and the Dead Sea to which it discharges, have experienced devastating pollution. When Palmer visited the Allenby Bridge[98] over the River Jordan he was able to cross the 'river' in just a single stride. The shrunken state of the river so shocked him that he wrote his newspaper article. In it he pleaded with all who value nature and the sacred to wake-up to what is happening to the river.

Palmer pointed out that the River Jordan is sacred to Christians, Jews and Muslims. For Christians it is the river in which Jesus was baptized. Although the exact place of his baptism is disputed, one possible site is Qasar al-Yahud, near Jericho in the West Bank[99]. Another location, on the Jordanian side, is Wadi Kharrar[100], also known as the Pools of Elijah where the prophet is reputed to have crossed the Jordan before being taken into heaven. This site is also sacred to Jews, and to Muslims, who recognise both Jesus and Elijah as prophets.

Palmer happens to be director of the Alliance of Religions and Conservation[101], which has been working with two other sacred and seriously polluted rivers, namely the Yamuna River in India and the Bagmati River in Nepal. He wrote that the Yamuna River flows sluggishly through the sacred city of Vrindavan, which is the birth place of Krishna, while the Bagmati River flows by the sacred Pashupatinath Temple in Khathmandu, where it is little more than an open sewer. Palmer reported that 'in both countries, religious organizations and local leaders have spearheaded programmes to protect and clean the rivers'. The partnership between religious and secular authorities is based on good economics and 'a vision of nature as a gift of God for which we have a responsibility to care'.

Today the more affluent developed countries put considerable effort into cleaning up their rivers. This is not just for aesthetic reasons, but far more importantly for the health of the environment and human beings. Uncollected and untreated wastewater poses serious health hazards through diseases such as typhoid and cholera. There is a range of other water-borne diseases that are prevalent in many developing countries through the discharge of various liquid wastes into natural and artificial streams. More developed countries have passed legislation to limit human exposure to such diseases. For example, the European Commission has approved the Urban Wastewater Treatment and Water Framework Directives[102]. Obligations are placed on all member states to provide effective treatment of wastewater from domestic and industrial use. Attention also has to be given to pollution from agricultural fertilizers and pesticides, as well as animal husbandry and other processes.

Many urban sewerage systems for wastewater and excess rainfall runoff (that is, stormwater) are combined. This can lead to large amounts of dilute sewage being discharged directly to the environment during significant storms through what are called combined sewer overflows. This happens because the treatment works are usually designed for only three times the average wastewater flow: it is unrealistic and uneconomic to treat much larger storm flows. Great efforts have been made in recent years to reduce overflow discharges by increasing the storage of sewage in the sewerage system and to make more effective use of treatment over longer periods of time. Real time control of the storage facilities (tanks and oversized pipes) is being introduced to improve the performance of the sewerage systems still further. But overflows still operate, and drainage engineers have to be on their guard to ensure that their systems perform consistently and sustainably[103].

The Arabic word for Islamic Law 'Shari`ah' is closely related to water. Its original meaning was 'the place from which one descends to water.' Before Islam appeared, shari`ah was a series of rules about water use, especially drinking water for all species. This good practice has however, lapsed over the last century. More attention is now being given to the need for clean water, and ancient practices are being restored. A number of Christian churches around the world are also waking up to their responsibilities to respect water and sacred springs. Ian Bradley, in his perceptive book 'Water: a spiritual history'[104], looks in detail at the way in which water has been regarded as a spiritual resource by different religions and communities[105]. Beginning with classical, Celtic and early Christian attitudes to water, he considers medieval holy wells, the Protestant promotion of spas, the cult of cleanliness, and the rediscovery of holy water in the late nineteenth century. There are many cases where springs are associated with healing, such as the Catholic shrine at Lourdes. In North America during the twentieth century, Catholic initiatives were focused on protecting and cleaning rivers such as the Hudson and Columbia River[106]. During the last part of the twentieth century there were considerable opportunities for churches to get involved with environmental groups and charities, such as A Rocha and Water Aid, to raise the awareness of the seriousness of pollution in our rivers and the urgent need for action to preserve our environment for future generations.

The desire to improve our management of wastewater from domestic and industrial use has prompted extensive research and entrepreneurial activity on the part of civil and public health engineers. They are searching for ideas to separate different forms of wastewater at source, and where possible to collect and treat it so that the resulting liquor can be returned safely to the environment, possibly with benefit of extracted raw materials for agriculture or other industries[107]. Other research focuses on naturally occurring or artificially constructed wetlands, containing particular plants such as papyrus, which are very proficient at removing nutrients from the water. These nutrients include nitrates and phosphates produced by agriculture, and toxic heavy metals that enter the water cycle through the disposal of industrial effluents or the wash-off of rubber and other materials from urban surfaces by rainfall[108]. Such modern forms of treating wastewater are still beyond the typical developing country, especially in rural areas. These countries are however showing great ingenuity in finding ways to improve the quality of drinking water, especially using cheap and readily available technologies.

References:

Reed beds: http://en.wikipedia.org/wiki/Reed_bed

M. Palmer (2010), the Guardian 27 July 2010

A-Rocha: http://www.arocha.org/gb-en/index.html

WaterAid: http://www.wateraid.org/uk/

Bradley, I (2012) *Water: a spiritual history.* Bloomsbury.

Resource questions:

How can we reduce the cost of treating used water in order to return it safely to the environment?

What can you do as a domestic user of potable water to minimize the pollution in your used water?

How would you argue that those who live in a temperate climate with more than 1000 mm of rainfall a year should take reasonable care of every litre of water they use?

18

As the waters cover the sea

'Deep calls to deep in the roar of your waterfalls; all your waves and breakers have swept over me. By day the LORD directs his love, at night his song is with me.'

Psalm 42:7-8

Read Isaiah 11:6-9

The Old Testament consists of 39 books that tell the history of one particular group of people, namely the Children of Israel, and their relationship to God. The books were written at various times for different purposes. Some books tell stories of how things came to be; others are written as history, others give instructions for life and worship of God, worship, wisdom, and prophecy. The books reflect the culture of the times in which they were written, the outlooks of the different authors, and the developing relationship between the Children of Israel and God. The authors (and editors) tell things as they are. They do not glamorize the behaviour of the people and their leaders. Instead, they strain for analogies to describe the often turbulent and tenuous nature of the relationship between God and his people. It is not therefore surprising that the Bible is full of analogies and sayings derived from nature that are used to illustrate human values, attitudes and beliefs and the interactions between individuals and God and people with each other.

Scattered throughout the wisdom and prophetic writings of the Old Testament are passages that link the nature and actions of God with water in one form or another. The Psalms and the prophetic books seem determined to remind us that our concept of God is far too small. They take the most frightening experiences that nature has to offer us and then state that God is so much greater, so much bigger than these awesome forces over which we have absolutely no control. A particular example that occurs frequently is that of the waves of the sea whipped up by the wind and crashing in the violent storm (Psalm 93:4). But even if these forces are beyond the control of man, God is always in control: 'the voice of God is over the waters' and 'The LORD sits enthroned over the flood' (Psalm 29:3, 10). And again: 'the LORD stills the waves of the surging sea' (Psalm 89:9) Indeed, the psalmist reflects that all creation should resound with the praise of God because the fact is that 'the LORD reigns' (Psalm 96:6, 9). But it is not only the psalms which speak of God being in control of nature; the prophets also envisage God controlling the waters and the hydrological cycle: 'When he thunders, the waters in the heavens roar; he makes clouds rise from the ends of the earth. He sends lightning with the rain and brings out the wind from his storehouses.' (Jeremiah 10:13) Habakkuk queries God's purposes when violent storms that afflicted his country: 'Were you angry with the rivers, O LORD? Was your wrath against the streams? Did you rage against the sea when you rode with your horses and your victorious chariots? . . . You split the earth with rivers; the mountains saw you and writhed. Torrents of water swept by; the deep roared and lifted its waves on high.' (Habakkuk 3:8). There is an assumption in the writings of the Bible that there is a permanence in the hydrology of the planet. For example, the short-term faithlessness of the people of Israel is contrasted with the 'permanent' snow on the mountains of Lebanon and the streams that are generated by snowmelt (Jeremiah 18:13)

I believe God created the seas and the floods, and he demonstrated his power time and again in the course of bringing his people to the Promised Land. God was at work in Noah's great flood, in the crossing of the Red Sea with the Egyptian army in hot pursuit, the crossing of the River Jordan to enter the Promised Land, and in the demonstration of his power to Job and others who were seeking the true nature of God. Human beings have much to learn when searching for God, especially when they start from a perspective that puts them at the centre of their universe. God challenges the point of view of those who are truly seeking him by asking questions: 'Who has measured the waters in the hollow of his hand, or with the breadth of his hand marked off the heavens? Who has held the dust of the earth in a basket, or weighed the mountains on the scales and the hills in a balance?' (Isaiah 40:12). He makes many promises through his prophets about what he will do for his chosen people in reordering the environment: 'I will make rivers flow on barren heights, and springs within the valleys. I will turn the desert into pools of water, and the parched ground into springs.' (Isaiah 41:18). In other words, the weather and climate are in his hands. We might bring about devastation on the earth, but God is in the business of bringing about his purposes. 'Water will gush forth in the wilderness and streams in the desert. The burning sand will become a pool, the thirsty ground bubbling springs. In the haunts where jackals once lay, grass and reeds and papyrus will grow' (Isaiah 42:6, 7). In his mercy and grace, God will be with his people, not necessarily to remove them from trials and difficulties, but to bring them through to a better place as he had done when he brought Israel out of Egypt, leading them through the wilderness, and encouraging them to fight to take possession of the land he had promised them. In all these things God was with them. Their part was to depend on him: 'When you pass through the waters, I will be with you; and when you pass through the rivers, they will not sweep over you. When you walk through the fire, you will not be burned; the flames will not set you ablaze'; see Isaiah 43:2. 'For I will pour water on the thirsty land, and streams on the dry ground; I will pour out my Spirit on your offspring, and my blessing on your descendants. They will spring up like grass in a meadow, like poplar trees by flowing streams.'

Our memories are short; we live largely for the present, forgetting the good things that have happened in the past and, if anything, remembering the bad things. When God remembers, it is to our benefit: 'To me this is like the days of Noah, when I swore that the waters of Noah would never again cover the earth. So now I have sworn not to be angry with you, never to rebuke you again.' (Isaiah 54:9). No matter how serious the situation, God promised to bring his people through, not just for their benefit but for the sake of his reputation and the glory of his name: 'Who sent his glorious arm of power to be at Moses' right hand, who divided the waters before them, to gain for himself everlasting renown, who led them through the depths? Like a horse in open country, they did not stumble; like cattle that go down to the plain, they were given rest by the Spirit of God. This is how you guided your people to make for yourself a glorious name.' (Isaiah 63:12-14) The God who led his people did so in order that 'They did not thirst when he led them through the deserts;

he made water flow for them from the rock; he split the rock and water gushed out.' (Isaiah 48:21). And if God did these things for them in the past, he will certainly do even greater things for them in the future: 'They will neither hunger nor thirst, nor will the desert heat or the sun beat upon them. He who has compassion on them will guide them and lead them beside springs of water.' In answer to those who still doubted that God could save them: 'When I came, why was there no one? When I called, why was there no one to answer? Was my arm too short to ransom you? Do I lack the strength to rescue you? By a mere rebuke I dry up the sea, I turn rivers into a desert; their fish rot for lack of water and die of thirst. I clothe the sky with darkness and make sackcloth its covering.' (Isaiah 50:2-3) Isaiah can but exclaim: 'Was it not you who dried up the sea, the waters of the great deep, who made a road in the depths of the sea so that the redeemed might cross over?' (Isaiah 51:10-11)

There is a strong tendency for us as individuals to think that God does not see into our hearts. As a consequence, we think we can impress God and others with our religious duties while our hearts are far from being devoted to him. As Jesus pointed out to his disciples, it is not what goes into us, such as food, that corrupts us, but what comes out of our hearts (Matthew 15:17-19). Our hearts determine our relationships with each other, whether founded on justice or injustice, righteousness or unrighteousness. The prophet Amos quotes God as saying that he will not listen to the music and words of our songs of praise. Instead, if we are to get God's ear we are to 'let justice roll on like a river, righteousness like a never-failing stream' (Amos 5:14). If you have seen one of the world's major rivers, such as the Nile or the Mississippi, or the Ganges or the Yellow River, or even a smaller national river, you can appreciate how we are to embed justice and righteousness into our governance structures and our personal dealings with each other as a permanent expression of our desire to be in harmony with the way God wants us to live.

There has been a strong and dramatic decline in church attendance in Europe in recent years: by the year 2020 forecasters predict that only 4.4% of the population of the UK will be attending church[109]. This is a decline of 55% over the forty years between 1980 and 2020. A consequence of this statistic is that there will be a growing ignorance of God's works and ways among the population.

This increasing ignorance is in contrast to the growing scientific understanding of the universe, such as the discovery of the Higg's boson which is claimed to be the fundamental building block of matter in the universe[110]. There is a deep confidence on the part of many that we are the masters of our own destiny, and religion is removed from the public to the private domain. Like Laplace, many say we have no need of God as a hypothesis to explain how the universe works, and therefore how society behaves. Even though we are unable to answer questions about the 'why?' of the universe within this knowledge framework, we are like the people in the story of the tower of Babel, who are confident that through the integrating power of the Internet and facilities such as Facebook®, we can build a real future where nothing is impossible for us. This self-belief is so strong that it refuses to acknowledge other contrary evidence of the deep and profound recession that has affected much of the (western) developed world, putting enormous strain on the euro—the European Union's single currency. Many comfort themselves with the view that such a recession is cyclic, but there is also strong evidence that it is due to significant greed and corruption at the heart of the banking system, compounded by avarice in the population as individuals seek to increase their wealth. European Christians need to have a better perspective of people's response to the gospel in other parts of the world, especially in East Asia, Africa and South America. There are reputedly far more Christians in China than there are in Europe, and evangelistic initiatives are more likely to come from Christians in those countries where persecution is more explicit than in Europe or North America. Whereas Europe has offered much to the rest of the world during the last millennium, it would appear that other countries and regions are now coming to the fore. European Christians may need to accept help gratefully from these 'overseas' Christians in seeking to re-evangelise Europe. What continues to be necessary is that justice must be seen to prevail, especially at the heart of the EU. As we quoted Amos above: we need 'Justice flowing like a river; righteousness like an ever flowing stream'. Another quotation, this time from Habakkuk and Isaiah, is that 'the earth will be filled with the knowledge of the glory of the LORD as the waters cover the sea' (Habakkuk 2:14 and Isaiah 11:9). May that day come soon.

Resource questions:

What other interesting and meaningful sayings concerning water have you come across in the bible?

Which biblical saying (involving water) has had the biggest impact on your life?

19

All that live in the water

'All earth's full rivers cannot fill
The sea that drinking thirsteth still'
Christina Georgina Rossetti from 'By the Sea' in 'Poems
and Prose', Oxford University Press (2008)

'The oceans are the planet's last great living wilderness,
man's only remaining frontier on earth, and perhaps his
last chance to produce himself a rational species.'
John L. Cullney, Wilderness Conservation,
September-October 1990

Read Psalm 107

The sea was disturbing to the Israelites. It was uncontrollably capricious. The storms affecting it were frightening in themselves. The high winds produced huge waves on the surface of the deep ocean and, although most sailors were unaware of it, the low atmospheric pressures at the centre of the storms helped to raise the sea level marginally below the storm centres. (The sea is raised by about one centimetre for each decrease of one millibar in the air pressure). The combination of high winds and low pressures resulted in significant surges along coastlines, similar to the tides induced by the moon and the sun. Sailors were respected by others because of their courage and their preparedness to face forces beyond most people's worst nightmares. The seas were used as a highway to bring exotic goods from faraway places, as well as the regular items of trade. The ocean was a source of food, namely fish, and in it lived creatures that triggered the fertile imagination of the sailors. Coastal waters were inhabited by salt water crocodiles, which, like their River Nile counterparts, were well capable of defending themselves and even attacking people. They crocodiles were respected and honoured by the local communities. The deeper ocean water was home to creatures of considerable size and frightening aspect: sharks, whales and sea serpents. Even today, exploration of the deep ocean trenches is revealing huge life forms that enhance our knowledge and imagination. Besides the crocodile, the Jewish writers were entranced by two particular creatures: the leviathan and the behemoth.

The leviathan is mentioned six times in the Old Testament. A detailed description of the beast is given by God in Job 41. The leviathan is the king of beasts. 'Nothing on earth is his equal—a creature without fear' (Job 41:33). Nobody can subdue him. No-one can rouse him. God made him, and who can stand against God? Because everything under heaven belongs to God, and the leviathan is beyond human control, then God must have made him for himself. The leviathan does not come under the stewardship that man has over all that moves on the land and in the sea; he is outside the charge that God gave to Adam in the Garden of Eden; see Genesis 1:26. The strength and gracefulness of his form, the fearsome teeth in his mouth, the scales tightly sealed on his back, the smoke from his nostrils, the firmness of his flesh and the hardness of his chest, all give him a terrifying form. Nothing that men can do to attack him with weapons can affect him. He leaves a trail in the mud like a threshing sledge; he churns the depths like a cauldron and stirs the sea like a pot of ointment. The Leviathan is invincible; only God can conquer it. Isaiah talks about God slaying the Leviathan, the monster of the deep, with his 'fierce, great and powerful sword'. It will be such a memorable day because then God will deliver Israel from Exile; see Isaiah 27:1. The Psalmist, in referring to God's invincible power to save, acknowledges that it was God who split the sea by his power, and broke the heads of the monster in the waters. He then identifies this monster as the leviathan, which ended up as food for the creatures of the desert; see Psalm 74:13-14.

So what was the leviathan? It is not clear whether it was some form of giant crocodile, or a sea-serpent or other form of beast left over from the age of dinosaurs and which is now extinct.

It is also difficult to identify the behemoth with any living animal. Some think it could have been an extinct dinosaur, while elsewhere in history the behemoth has been associated with the hippopotamus, the elephant and other large animals. However, whatever its identity, it is a beast that like the leviathan cannot be tamed by man. In effect, it is a creature of Yahweh, a being outside the control of human beings, just as the weather is beyond our control (Ecclesiastes 11:5).

One other sea-based animal has an important mention in the Old Testament, and that is the sea cow. Its hide was particularly valued by the Israelites as a means of wrapping up the delicate vessels of the tabernale used in the various rituals. This was done when the people were about to set out on their travels in the wilderness; see Numbers 4:6.

There are of course many other amazing ocean creatures that we are familiar with today. Most children taught Biblical stories have heard about Jonah and the whale. In the Book of Jonah we learn that God was concerned about the great wickedness of the inhabitants of the Assyrian capital city of Nineveh. He ordered Jonah to go and prophesy against the city. But Jonah did not appreciate God's call because he knew that if he went and prophesied to the people, they would repent, and that nothing of the punishment Jonah proclaimed would happen. Consequently he, Jonah the prophet would appear foolish, because what he had prophesied would not happen. And all this because he knew God was full of compassion for human beings. So Jonah had other ideas. He tried to run away from God by taking passage on a ship going from Jaffa to Tarshish, which is in completely the opposite direction to Nineveh. In other words, he was determined to put as much distance between himself and Nineveh as he could.

But Jonah had not reckoned with what God would do. A huge storm arose. The sailors realized it was no ordinary storm. Somebody must be to blame for it. So they cast lots and discovered that Jonah was responsible. When challenged, Jonah admitted it was his fault. He stated that the storm would cease if he was thrown overboard. This was the last thing the sailors wanted to do, and they tried desperately to get to shore. When they failed to do so, they realized they had no option but to do what Jonah had said; unwittingly as it turns out this was God's intention all along (Jonah 2:3). Once Jonah was thrown over the side of the ship, the sea became calm. That was the satisfying end of the story for the sailors, but for Jonah it was still only the beginning, for God had made plans for Jonah to be saved by a huge fish (or whale) that swallowed him whole. For three days and nights in the fish (Jonah 1:17) (that is, as long as the time Jesus was in the tomb) Jonah had time to think over what had happened. He prayed to

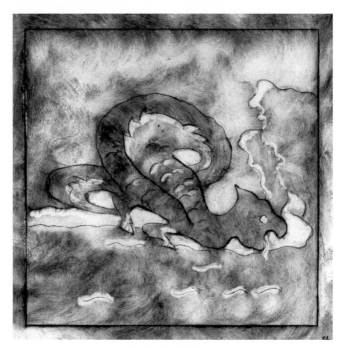

The Leviathan

God, and turned from any thought of recriminations to thanksgiving, vowing to do what God wanted. So God commanded the fish to vomit Jonah out onto dry land.

Again God ordered Jonah to go to Nineveh and to proclaim that Nineveh would be overthrown in forty days' time. Jonah was obedient to God this time. Perhaps not to his surprise, the people of Nineveh believed his word. No less a person than the king of Nineveh put on sackcloth and sat in ashes. He made a proclamation decreeing a fast with prayer and repentance. God saw the response of all the people, and spared the city.

God still had more to teach his reluctant prophet. You can read about Jonah in the book called by his name. When asked for a miraculous sign by the Pharisees and teachers of the Law, Jesus referred them to this book. He stated that no sign would be given to that generation but the sign of Jonah, who after three days and nights inside the great whale was released back to life. This prefigured Jesus being in the tomb for the same length of time, after which he would be resurrected to new life. Jesus added that at the judgment, the people of Nineveh would stand up and condemn his generation, because the people of that great city repented when hearing Jonah's proclamation, and now somebody far greater than Jonah was there; but though Jesus was greater than Jonah, his hearers did not listen to him; see Matthew 12:39-41. Even in their ignorance they were condemned.

Resource questions:

How can we better sustain species of large animals in danger of extinction?

Are you aware of what God has called you to be and do? How can you be obedient to God's call on your life?

20

Water and character formation

'Jack and Jill went up the hill to fetch a pail of water;
Jack fell down and broke his crown, and Jill came
tumbling after.
Then up got Jack and home did trot as fast as he could
caper,
To old Dame Dob who patched his nob with vinegar and
brown paper.'

Old nursery rhyme

'Smooth runs the water where the brook is deep'.

William Shakespeare (1564-1616), Henry VI, part II

'There is nothing softer and weaker than water,
And yet there is nothing better for attacking hard and
strong things.
For this reason there is no substitute for it'.

Lao-Tzu (c. B.C. 550),
See en.wikipedia.org/wiki/**Laozi**

Read Isaiah 43:16-21

As we have seen, water is a weird, unique substance; and because it appears in our environment in so many different forms and circumstances we can learn a lot from it. It may be inappropriate to anthropomorphise water, but De Mello (1989) uses this approach to explain what he means by 'enlightenment'. He suggests that an enlightened person is like water; that is, he or she 'is soft and pliant to the touch and yet irresistible in its power. Water does not strive in competition with other things, but benefits all beings. It is selfless and does not have an ego to be satisfied, yet through it and by it others are transformed. As a result of its presence the whole world prospers. It does not have desires that contend with others, who are therefore left unspoiled. It is indifferent to its location; there is no preferred place of status. It exists and acts sweetly and powerfully. It fulfils its destiny.' By replacing 'it' with 'I' De Mello would encourage us to have an attitude in which we are enlightened in our being and existence both individually and corporately.

It is no surprise that images of water are often used to explore both human and divine behaviour in the Bible. This may be because the biblical writers, in their largely arid environment, recognised how important water is as a resource. You only have to glance through the Psalms or Isaiah to see how frequently water is referred to in some form or other: clouds, thunder, snow, rain, mists, dew, brooks, streams, rivers, floods and droughts are used as illustrations of human behaviour or of God's concerns. Facts about the physical world of water are translated into reflections on the human condition. For example, the undisturbed surface of water gives a true reflection of images, including the reflection of the face of anyone who looks into it. Proverbs picks up this fact and deduces that 'as water reflects a face, so a man's heart reflects the man' (Proverbs 27:19). That is to say, the heart reveals the nature of a person, as Jesus pointed out so radically to his disciples (Mark 7:20-22). The timing of natural events involving water is used by the writer of Proverbs to highlight the nature of what he calls 'the fool'. Such a person does not deserve honour. To give honour to the fool is as incompatible as snow in summer (in the northern hemisphere) or rain during the harvest; see Proverbs 26:1. In this chapter, we look particularly at the allusions to water in the wisdom literature, namely, the books of Job, Proverbs and Ecclesiastes.

The book of Job is a study of human suffering and the sustainability of faith in God in the context of this world. Job suffers the loss of virtually everything, including his property, family and health, at the hand of satan, and yet he retains his faith in God. He is attended by four 'comforters', who advise him on his understanding of himself and his relationship to God. In the conversations, Job and his comforters make frequent use of illustrations involving water. For example, Eliphaz berates Job for underestimating God's wonders: 'He bestows rain on the earth, he sends water upon the countryside' (Job 5:10). Job responds by saying that a despairing man would normally have the support of his friends, but his 'brothers (the comforters) are as undependable as intermittent streams, as the streams that overflow when darkened by thawing ice and swollen with melting snow, and in the heat vanish from their channel' (Job 6:15,16).

Bildad, another comforter, attacks Job with an accusation as common today as it was then. Job had failed to couple his present demise with his sin: 'Can papyrus grow tall where there is no marsh? Can reeds thrive without water?' In other words, is it not obvious that Job's awful state is a consequence of his sin? But Job, far from seeking to justify his righteousness, responds that no mortal can be righteous before God. It is not a matter of what Job has done specifically to get himself into his present situation, because in general, like his comforters, he is a sinner. And so no-one can resist God: 'He alone stretches out the heavens and treads on the waves of the sea' (Job 10:8). So Job insists on his innocence in the face of continued accusations, despite the fact that God caused his 'forces (to) come against me wave upon wave' (Job 10:17). In response to yet further accusations on his integrity, this time by Zophar, Job acknowledges God's wisdom and power: 'If he holds back the waters, there is drought; if he lets them loose, they devastate the land' (Job 12:15). It is God who takes the initiative. It could be that 'a man lies down and does not rise' because he sleeps in death: life has drained from him; see Job 14:12. The parallel in nature is 'As water disappears from the sea or a river bed becomes parched and dry' (Job 14:11) so the sea or river loses its water and dies.

Job is frustrated because he says that he searches for God to present his case, but he cannot find him. Others, it seems, deliberately set out to avoid God; they are active in the darkness, breaking into houses at night, and hiding during the daylight hours, 'shutting themselves in'. Yet, Job sees them as 'foam on the water' (Job 24:18), generated by some pollutant in the turbulent flow separated from the surface of the water when the wind blows. In other words, God will judge them and they will have no defence.

Job is profoundly perceptive of the wisdom of God expressed in creation. He proclaims God's amazing creation: 'He wraps up the waters in his clouds, yet the clouds do not burst under their weight' (Job 26:8). We simply do not know how the weather will perform in the longer term because any forecast is so sensitive to the initial conditions pertaining at the time. Real wisdom, says Job, comes from the fear of God (Job 28:28), and he illustrates this by noting that God 'measured out the waters' at creation (Job 28:25); God knew in detail what he created. Job then reflects how he had lived his life in the time before God brought him to this state; 'My roots will reach to the water, and dew will lie all night on my branches' (Job 29:19). Elihu interrupts by saying that Job still lacks awareness of how great God is: 'He draws up the drops of water, which distil as rain to the streams, the clouds pour down their moisture and abundant showers fall on mankind. Who can understand how he spreads out the clouds, how he thunders from his pavilion? See how he scatters his lightning about him, bathing the depths of the sea' (Job 27-30).

At long last, God speaks to Job. God reflects on how little Job really knows about the amazing scope and extent of creation, and what God has done in it. God challenges Job: 'Where were you when I laid the earth's foundation?' (Job 38:4). 'Who shut up the sea behind doors when it burst forth from the womb, when I made the clouds its garment and wrapped it in thick darkness, when I fixed limits for it and set its doors and bars in place, when I said, 'This far you may come and no farther; here is where your proud waves halt'?' (Job 38:8-10). He continues to ask Job: 'Have you journeyed to the springs of the sea or walked in the recesses of the deep?' (Job 28:16). 'Have you entered the storehouses of the snow or seen the storehouses of the hail . . . ?' (Job 28:22). God asks Job: 'Who cuts a channel for the torrents of rain, and a path for the thunderstorm, to water a land where no man lives, a desert with no-one in it, to satisfy a desolate wasteland and make it sprout with grass? Does the rain have a father? Who fathers the drops of dew? From whose womb comes the ice? Who gives birth to the frost from heavens when the waters become hard as stone, when the surface of the deep is frozen?' (Job 38:25-30). 'Can you raise your voice to the clouds and cover yourself with a flood of water?' (Job 38:34) and 'Who can tip over the water jars of the heavens?' (Job 38:37). Faced with such an overwhelming catalogue of God's acts, Job can only fall down and confess his nothingness in God's sight, and how he had presumed on his favour. Whereas God talks about much more than these gigantic themes of creation involving water, they nevertheless highlight the place of water in creation and in our lives. We do well to ponder the greatness of God when we are tempted to think how great our own achievements are.

The wisdom literature in Proverbs and Ecclesiastes contains a number of sayings involving water in some form or other. For example, 'Many waters cannot quench love' (Song of Solomon 8:7). The implication is that although we use water to douse an intense fire, the burning ardour of love is beyond its extinguishing power. Again, the saying 'Drink water from your own cistern, running water from your own well' concerns the married man being faithful to his own wife and not seeking pleasure with someone else. In other words, 'Should your springs overflow in the streets, your streams of water in the public squares? Let them be yours alone, never to be shared with strangers. May your fountain be blessed, and may you rejoice in the wife of your youth' (Proverbs 9:15-18). Then there is the recognition that the woman who declares that 'stolen water is sweet' (Proverbs 9:17) is leading a man into promiscuity. On the other hand, 'a quarrelsome wife is like a constant dripping on a rainy day' (Proverbs 27:15). Strong warnings indeed!

A number of proverbs refer to relationships between people: family, friends, neighbours, colleagues, strangers and even enemies. When water finds a path to leak through a dam, in time it can wear away the material in the crack so that it progressively enlarges it. Unless the crack is noticed and repaired, water flows through it unrestrained and the whole structure can collapse with disastrous consequences. The time to deal with the potential collapse of the dam is right at the beginning when the crack is small, because a repair is then comparatively straight forward. So the advice of Proverbs is to drop any matter that is leading to a quarrel before a dispute breaks out! (Proverbs 17:14)

Dealing with an enemy is very difficult for many people because it is hard to be sufficiently emotionally detached to approach the person (or people) positively. Proverbs 26:21-22 advises us that 'If your enemy is hungry, give him food to eat; if he is thirsty, give him water to drink. In doing this, you will heap burning coals on his head.' Paul, probably quoting Jesus, strongly advocates that we consistently follow this advice (Romans 12:20), not least because it reflects Paul's own attitude to his fellow human beings.

On the other hand a righteous man who gives way to the wicked is 'Like a muddied spring or a polluted well' (Proverbs 25:26). The spring is normally a vital resource, but when its water is unusable it becomes more of a nuisance than a benefit.

There are a number of sayings referring to the way people communicate with others through words or the lack of them. The emphasis is on the use of few words to convey profound insights, rather than many words in which there is little content. For example, 'the words of a man's mouth are deep waters, but the fountain of wisdom is a babbling brook' (Proverbs 18:4). Here it could mean that a person talking from his own need may find it difficult to explain what he is feeling, whereas if called on to give words of wisdom he becomes eloquent and is able to explain things in simple terms. 'Deep in water are the purposes of human hearts, the discerning man has only to draw them out' (Proverbs 20:5) is a way of saying that a person has hidden depths that are not immediately apparent and may require particular circumstances for these depths to be exposed. On the other hand, people can be categorized as being 'tossed to and fro like the waves' (Ephesians 4:16). Here Paul is identifying a certain group of people who appear to be unable to be at rest in their religious thinking, continually changing their minds, being carried along by 'every wind of doctrine'. Other people, however, express a deep desire and longing to know God better, 'As a deer longs for a stream of cool water' (Psalm 42:1). Such people open themselves to the experience of joy—so it is with 'joy that we draw water from the wells of salvation'; see Isaiah 12:3; 30:30; 32:3; and 49:10.

Finally, in Ecclesiastes 11:1 we are invited 'to cast bread upon the waters'. This enigmatic saying calls us to engage in life, and not to be paralyzed by our inability to find meaning in the totality of our lives. The saying appears to invite us to be involved in commercial enterprises involving overseas trade (see 1 Kings 9:26-28; 10:22). Indeed, Ecclesiastes 11:1-6 says much about diversity in investments, that is, not putting 'all your eggs in one basket', or not taking so much notice of prevailing circumstances that they prevent you from taking action when you should, and therefore taking reasonable risks. Life is a risk, but let us act and enjoy what comes while committing what happens in the future to the God who loves us.

Reference:

Mello, A de (1989) *The Heart of the Enlightened*, Harpers Collins, London

Resource questions:

There is much wisdom in the different books of the bible. What biblical principles do you consciously apply to your job or career?

How can you improve your knowledge of God, his works and his ways?

Use a concordance to look up references to 'water' as they appear in the Bible. What other insights can you find about God and his creation, and the activities, ambitions and motivations of people?

21

Dew and grace

'Rain is grace; rain is the sky condescending to the earth;
without rain, there would be no life.'
> John Updike, Self-Consciousness: Memoirs, 1989

'He loads the clouds with moisture;
he scatters his lightning through them.
At his direction they swirl around
over the face of the whole earth
to do whatever he commands them.
He brings the clouds to punish men,
or to water his earth and show his love.'
> Elihu in Job 37:11-13

Read Judges 6:33-7:25

I remember being in a boat on a Dutch canal on the outskirts of Leiden one sunny autumn day. As we cruised slowly back from our excursion into the depths of the countryside, we were aware of the air temperature falling. The water level in the canal was above the adjacent land, and we could see down onto fields that shimmered in the fading light. The fields were covered with spiders' webs on which dew had condensed. It was as though millions of tiny jewels on the grass were reflecting the dying sun, producing a mystical, entrancing experience.

The phenomenon of dew occurs when water changes from a gaseous or vapour state to a liquid state. Water vapour makes the air humid. If this air is cooled below a certain temperature, then, depending on the barometric pressure, the water condenses and collects on any available surface. During that boat trip, the moist air near the ground cooled first, and each spider's web provided a surface on which the condensing water could collect, in definite drops like rows of beads on a necklace. Since the thread of a spider's web is comparatively strong, it is able to hold the weight of the condensed water. Sometimes, water droplets can be seen on the leaves of plants on a cool morning. In such a case, water vapour is emitted onto the surface of the leaves by the plants through the process of transpiration. Normally this water evaporates immediately, but due to the coolness of the air, it collects and remains on the leaves before the water can escape. Again, we might experience dew on a chilly morning when water condenses on a car windscreen. It is like looking through a mist, and the screen must be wiped clean before driving. If the temperature falls below 0°C it is likely that the condensed water or dew will freeze, creating ice on the car windows. The amount of water collected on the surface is usually small. Once the sun comes up, its radiation will melt the ice or evaporate any dew. This phenomenon prompted the prophet Hosea's caustic remark that Israel's love is like the morning dew—it vanishes when the sun comes up! (Hosea 6:4). But the Psalmist used the concept of dew in a more encouraging way when expressing the importance of brothers living together in unity; he pointed to the experience of dew falling over Mount Hermon as a sign of God's blessing on his people; see Psalm 133:3. Elsewhere in the Bible, Micah uses dew as an analogy for the remnant of Israel which will become a means of gently and consistently distributing God's blessing to many nations (Micah 5:7).

The writer of Proverbs is well aware that when a king is angry, everybody knows about it, but when the king expresses his favour it is like the dew, and everyone benefits (Proverbs 19:12). Then again, when God is encouraging his chosen King to sit at his right hand until his enemies submit to his rule, young men will volunteer for his army 'like the dew of the early morning'; Psalm 110:3). Dew is distributed across all surfaces exposed to the cool air, so God's grace and blessings are distributed to all without distinction.

The tendency of cool mornings to generate conditions for dew to form, even in desert regions before the sun comes up, has long been utilized by experienced travellers to collect some of this condensed water by getting it to run off the underside of a surface such as an inverted cone. Water collects on the surface, runs down the underside of the cone and drips off the tip into a water tight vessel. Given the right conditions and size of cone, enough water can be collected to keep a person alive even in the most rigorous conditions.

One person who was familiar with dew was Gideon. He was a reluctant Judge of Israel, that is, an arbitrator-leader appointed by God to rally the tribes of Israel during the early stage of their occupation of the land of Canaan. The book of Judges is a record of how the Children of Israel continually turned away from following God and worshipped the idols of the nations around them. This always resulted in the Canaanite tribes oppressing and harassing the Israelites, who finally cried out to God for deliverance. After one such episode of apostasy, God called Gideon to lead the people against their enemy, and at the same time to condemn their worship of idols. But Gideon doubted his own ability and, like many reluctant leaders, was unsure of God's calling. God proceeded to develop Gideon's character. First, God told him to confront the apostasy of the people in his own home town. He was to destroy the altar of his home town dedicated to the foreign deity, Baal, and the symbol of the goddess, Asherah, beside it. He did this at night time rather than openly during the day in front of the people. When the townspeople found out, they wanted to kill him. Fortunately Gideon's father, Joash, defended his son.

Then Israel's enemies, the Midianites and Amalekites, assembled in the Valley of Jezreel. The Bible reports that 'the Spirit of the Lord took control of Gideon' (Judges 6:34). He rallied four of the tribes of Israel to join him in battle. But he was still unsure that God was going to use him to rescue Israel. So he decided to seek assurance that God really was calling him to such a drastic action by requesting God to provide him with a sign; see Judges 6.33-40.

His idea was to put a fleece of sheep's wool on the ground in the place where the community threshed its wheat. He challenged God to make the fleece wet during the night, while the ground around it remained dry. When he woke in the morning, the ground was dry and the fleece was soaked with dew; there was enough water to fill a bowl. But despite this being very extra-ordinary, Gideon was still not convinced. In order to make really sure, he wanted to test God one more time. So the next night he requested God to make the ground wet with dew while the fleece remained dry. And so it happened. The rest, as they say, is history! But not quite: Gideon was going to be involved again with water.

Gideon gathered a battle force of more than 30,000 men from the neighbouring tribes. The more men he had under his command the better. But God had a surprise for Gideon: He informed him that there were too many men in his army! God would save His people and fight for them, but he did not want the Israelites to have any reason to claim the victory as their own. So Gideon had to reduce the number of men in his force. But how? Which men should he send home? God told Gideon to let those men who were afraid of fighting go home. 22,000 men took up this opportunity, which left 10,000. This number was still too large for God. He instructed Gideon to take the men down to the water to drink. Once there, God told Gideon to separate those men who cupped the water into their hands, raising it to their mouths, from those who knelt down and directly lapped the water with their tongues. Three hundred men chose to do it the first way. It was with these few that Gideon was able to bring panic among the far greater number of enemy forces during the night, and subsequently to rout them. God used the few to destroy the many (Judges 7:17-22). And all because of dew and the way in which a man drinks water!

Resource questions:

How do you seek God's will and purpose in your life?

What do you think of the 'fleece' method of finding out what God thinks you should do?

What other ways are open to you to know God's will for you?

22

Water and God's word

'Be glad, O people of Zion,
rejoice in the LORD your God,
for he has given you
the autumn rains in righteousness.
He sends you abundant showers,
both autumn and spring rains, as before.'
                    The prophet Joel (c. 835 BC) in Joel 2:23

Read Is 55:6-13

We have repeatedly come across some radical aspects of the word of God-God speaking creation into being, communicating directly with human beings, stating his authoritative law and commandments through Moses to his people, and proclaiming his judgment and call through his prophets to Israel and all nations. In general, God's word is creative in bringing life and hope for those who seek him, but judgment and death for those who oppose him. Ultimately God speaks in and through the person of his Son Jesus Christ, whom the Apostle John refers to as the Word of God. Jesus came to reveal the Father. As the Word of God, Jesus is the complete expression of the Father and his will and purpose.

Isaiah was a keen observer of nature, and knowledgeable about the part played by water in the weather. In particular, he used his insights sensitively to highlight how God's word brings about its intended purpose. For Isaiah, rain and snow falling on the earth make a vital contribution to life on the land surface and beneath it. Such precipitation is a deliberate gift of God in the whole scheme of the hydrological cycle to foster the life of plants and animals, and especially human beings; see Isaiah 55:10-11. Rain provides the roots of plants with the water they need. This water is taken in by the roots and circulated to all cells in the plants. Because of this, plants can grow and reproduce themselves. When water is in short supply, plant life on the earth's land surface suffers. Animals and human beings, like plants, are largely restricted to the earth's surface even if they are mobile; and, like plants, they too depend on fresh water being available, if not directly from lakes, rivers or groundwater then indirectly from rainfall. If there is no such resource they suffer. Water therefore is a key to ensuring the sustainability of life on the earth.

We were created for fellowship with God. In order to sustain this fellowship we are dependent on God's communication with us, in terms that we can appreciate and understand. Isaiah sees God's word as being purposefully sent and given to all people, whether they acknowledge it or not, and that it will achieve the particular objectives that God intends; see Isaiah 55:10-11. However, for God's word to bear fruit in people's lives, it needs to be taken up by individuals, who are then responsible for allowing that word to influence their lives. They need to respond to the word, and even be transformed by it in ways similar to those in which plants benefit from the rain. The people who do this can expect to flourish, producing in their lives good deeds which correspond to 'sufficient seed for the sower to grow next year's crop and to mill the flour for the baker to make bread'. To draw on yet another analogy, such people will be like fruitful branches on a vine. Just as Jesus referred to himself as the true vine, and to the importance of each of his disciples being branches attached to the vine, this enables the pruning of the branches by God our Father through his word active in the disciples' hearts and lives, leading to much fruit the following 'season'; see John 15:1-10. The most important thing to note is that God's word always accomplishes what He intends: He cannot be denied.

We may be ignorant of how God achieves his purposes in us, just as we are unaware of how growth occurs in a plant. Job refers to people 'drinking in' God's words as the ground absorbs the spring rain (Job 29:23). Such words may be the result of God searching the sources of the 'rivers' in us, and bringing hidden things to light; see Job 28:11. Again, God's words are like the seed sown on good and poor soil; his rain falls on the just and the unjust, so all share in his grace. In contrast, there may be a famine of 'hearing the words of God' (Amos 8:11), the implication being that we are dependent on regularly hearing God's word to us, and if this does not happen we suffer. We are like the land that is never satisfied by having enough rainfall; see Proverbs

30:16. Similarly, God may open the 'floodgates of heaven' in response to his people's acceptance of him (Malachi 3:11)—or not opening them, if they reject him.

It took Jesus' disciples time to come to terms with who their Master truly was. Indeed, it was only through personal revelation from Jesus and the Father that they came to know the truth about Jesus. On one occasion, Jesus revealed himself in all his glory to Peter, James and John by being transfigured before them with Moses and Elijah on the top of a mountain; see Luke 9:28-36. Following Peter's vain attempt to detain Moses and Elijah with Jesus, by offering to build three tents for them, God spoke from the cloud that overshadowed the disciples, telling them that Jesus is the one they should ultimately listen to, compared with Moses (who had given the law to Israel and represented the authority of God) and Elijah (who was the most famous of the prophets and spoke with the anointing of the Holy Spirit). Jesus, who had both the authority of God and was anointed by the Holy Spirit for his earthly ministry, is the unique human being who reveals the Father. God the Father resides in the Son who is at one with Him, doing what he sees his Father doing and saying only what the Father says. And Jesus has the words of eternal life. His teaching, and especially the parables he told, contained profound truths about the kingdom of God as a present reality, in memorable stories that could be told and retold in widely different contexts.

To conclude this chapter, we consider two such well-known parables, namely (i) the sower and the seed and (ii) the man who built his house on the rock. The first parable does not mention water explicitly, but of course it is necessary for the seed to germinate and grow, no matter in what soil the seed is sown. The sower casts his seed uniformly over the land, indicating that God's grace is extended to all, just as the rain that provides moisture to the ground is distributed uniformly over a field. The seed however falls on different surfaces: a path, rocky soil, ground with weeds, and good soil. The seed on the path is picked up by birds. The seed on rocky soil starts to grow well, but the roots find it difficult to penetrate between the rocks to find moisture, and the result is that the tender shoots wither and die. The seed falling on soil covered with weeds has to compete with the weeds for what little moisture is present; and because the weeds are acclimatized to the conditions and aggressively seek out what moisture exists, the seed is unable to compete with them for water and light. The young growth is choked by the weeds, and the seed cannot be fruitful. But the seed on good soil is able to grow well and to be fruitful, producing many new seeds for each single seed that was sown.

Jesus normally told his parables to the crowds around him, and then left his hearers to work out what the parables meant to them. Most of Jesus' parables can have a number of different interpretations in different contexts. But sometimes Jesus interpreted a parable for his disciples, as on this occasion. The seed, he said, is the word of God. The different types of ground represent people who receive God's word differently. The path represents those who are prepared only to consider God's word briefly, because it has little meaning for them. The rocky soil is a picture of those who initially find some meaning in God's word so that they want to explore it more: but as time passes, they fail to find it interesting or important enough to command their attention; the initiative is lost and God's word becomes ineffective in their lives. Similarly, the soil with weeds represents those who receive God's word in such a way that it takes root in their hearts, but their cares for what the world offers tend to choke the word; although it has some influence on their lives, it is prevented from bringing about real change. This leaves those who respond to God's word in such a way that it takes permanent root in their hearts, and bears fruit in their lives, dependent on how God has blessed them.

The second parable is very different. Here Jesus asks the question: Who is wise? One who builds his house on the rock or on the sand? For when the storm comes and the flood waters beat against the house, the one built on the rock will survive, while the other built on the sand is likely to collapse. This is because sand provides an unstable foundation, with a tendency to experience movement that threatens the structures built on them. (Here I have to admit to knowing that some sands form a particularly rigid foundation when compacted: this knowledge was used very effectively by Dutch engineers when building the Oosterschelde Barrier in the south of The Netherlands following the disastrous 1953 floods in the Zuid Holland area[111]. Also, our own house in The Hague was built on a dune of pure white sand over a hundred years ago, and it is still standing secure today!)

If rainfall reflects the activity of the word of God, it also refers to the blessing of God's Spirit. So God will pour out his Spirit on a thirsty people, just as water is poured on the thirsty land (Isaiah 44:3). This happened on the day of Pentecost at the birth of the church (Acts 2:1-4). Throughout Christian history, there have been outpourings of the Holy Spirit on groups of people seeking God's blessing (see Hosea 6:1-6), with many receiving particular gifts to promote the proclamation of the gospel and the announcement that the kingdom of God is present in this world now. An important characteristic of such a movement is joy (Isaiah 12:4) as God's people share in the risen life of Jesus.

Resource questions:

In what ways have you heard (or been aware) of God speaking to you?

How important is God's word (spoken, read, heard) in its different forms for you?

23

By the waters of Babylon

'I do not know much about gods; but I think that the river
Is a strong brown god . . . . almost forgotten
By the dwellers in cities . . . Unhonoured, unpropitiated
By worshippers of the machine, but waiting, watching
and waiting.'
'The Dry Salvages' in 'Four Quartets' by T. S. Eliot, first
published in 1944 by Faber & Faber Ltd, London

Read Psalm 137

Most western countries today host groups of people who are seeking asylum, that is, residence in a country not their own in order to avoid persecution, poverty or natural disaster back home. This is a voluntary form of exile. But in the eighth century BC, the Assyrians adopted a ruthless way of subjugating the peoples of the nations they conquered. They deported the inhabitants of one country *en mass*, and distributed them between several other countries under their control. So when they attacked and overcame the ten Northern tribes of Israel, the Assyrians forcibly moved most of the population elsewhere. This Exile was devastating for the people of Israel. They had been removed from the very land God had promised them: the land which their ancestors had conquered and in which they had permanently settled. Now God, it seemed, had gone back on his promise, and, uprooted from their land, they experienced a confusion of cultures as they were forced to adapt to new situations elsewhere. In the chaos of this uncertainty, the Assyrians maintained control over their conquered subjects.

The tribes of Benjamin and Judah escaped this first wave of deportations because, when Sennacharib, the Assyrian king was about to take Jerusalem, he was forced to retreat because of the mysterious deaths of 185,000 of his soldiers. He returned to his capital, Nineveh, where he was assassinated; see Isaiah 37:36-38. Judah continued to survive as an independent nation until the sixth century BC. In the meantime, the Babylonians became the dominant power in the Middle East. As they grew in strength they too attacked Jerusalem; in 586 BC they successfully took the city. Most of the remaining two tribes were also torn from the land God had promised them, and were taken into exile to Babylon. Jeremiah poetically describes the corporate mourning of his people through the imagined despair of Rachel, (Jacob's wife), when all the young men of noble birth were led into exile from Jerusalem to Babylon (Jeremiah 31:15); among them were Daniel and his three companions Shadrach, Meshach and Abednego. Some of the exiles, like Daniel, endured the reigns of successive kings in Babylon (Daniel 1-6). Like Joseph centuries before, Daniel's integrity and trust in God enabled him to rise to positions of great power under each subsequent king. Daniel lived sufficiently long to almost see out the first phase of the time in exile; but he does not appear to have ever returned to Jerusalem after the seventy years of exile that had been prophesied by Jeremiah were over.

Scholars think that much of the book of Daniel was written in the Maccabean period of Jewish history (second century BC), which was after Malachi, the last book in our version of the Old Testament, was written (probably in about 450 BC), and well before any of the New Testament books. The book of Daniel also contains visions of what God was doing and was going to do in the world; see Daniel 7-12. Many have tried to interpret these visions in terms of known world history. The visions are both challenging and frustrating; they contain many parallels with human history, but they do not fit in precisely with our understanding of it. The desire to interpret the visions precisely has led many astray, and has been the cause of not a few calamities in the Christian church.

Daniel was not the only prophet of the Exile; there were many others: Jeremiah, Ezekiel, the second author (Deutero-Isaiah) who contributed to the book of Isaiah, and various 'minor' prophets, including Amos, Zechariah and Micah. Ezekiel, like Daniel, received a number of visions while in exile, many of which were received beside the River Chebar, a tributary to the Euphrates (see Ezekiel 1:1, 3:15, 10:15 etc). Perhaps the gentle flowing river encouraged Ezekiel to reflect more deeply on the condition of the exiles and the longing of Yahweh for his people. His visions in particular appear to have involved transporting him physically to Jerusalem in order to be present at critical events in Jerusalem as they unfolded; see, for example, Ezekiel 11:1.

The exiles continually pined for Jerusalem. Nothing in their new land could remove the sadness and pain of being separated from their home Psalm 137 records their tears as they remembered what they had left behind. 'By the rivers of Babylon we sat down and wept when we remembered Zion.' The exiles congregated on the river banks to share their sorrows, and to weep deliberately for Jerusalem.

The first lines of Psalm 137 are very well known, and have even formed the lyrics of modern pop songs[112]. They describe the sadness of the Israelites, when asked by their captors to 'sing the LORD's song in a foreign land'. They refused to do this, leaving their harps hanging on the trees. If they did not sing, they nevertheless exhorted themselves to remember Jerusalem. The psalm ends with violent images full of revenge, in which a 'Daughter of Babylon' is told of the delight the author feels on envisaging 'he who seizes your infants and dashes them against the rocks', something that had been done to their own children. Rabbinical sources attributed the psalm to the prophet Jeremiah, and the Septuagint version bears the superscription: 'For David. By Jeremias, in the Captivity.'

Whether in Babylon, or earlier in Nineveh, the exiles from Judah and from the ten Northern tribes of Israel sought to keep the flame of hope alive through their use of song and telling repeatedly the stories of their history and traditions.

The sustainability and success of Assyria and Babylon depended on a regular supply of sufficient water for the main centres of population. This required considerable engineering and management skill. Sennacharib, King of Assyria, sought to improve the quality of the water from the Tigris delivered to his capital, Nineveh, and his palace at Khorsabad, by damming particular tributaries and bringing the water from the river by canals to a reservoir. Another larger, magnificent canal then brought the water to its destinations. The exiles were probably involved in the construction of this large canal, and therefore shared in its benefits for the city. Babylon was situated on the banks of the Euphrates, some 85 km south of present day Baghdad, in the region of Mesopotamia. It too had access to extensive supplies of fresh water.

By the waters of Babylon

A particular feat of water engineering that benefitted both Nineveh and Babylon were 'qanats'[113]. These were artificial underground channels that collected water from springs or water bearing strata, and conducted it to cities and regions of intensive agriculture. The design of the qanats originated in Armenia, and was perfected in Persia, especially during the reign of Darius (521-485 BC).

But one of the wonders of the ancient world was the hanging gardens of Babylon. These were thought to have been built by Nebuchadnezzar II around 600 BC. It appears that he ordered the building of the gardens in order to help his wife, Amytis of Media (Persia), stop pining for the trees and fragrant plants of her homeland. The hanging gardens are extensively described on tablets by the Greek historians Strabo and Diodorus Siculus; see Finkel (1988)[114]. What is remarkable about these tablets is that they describe the use of a mechanism for lifting water that is similar to the Archimedes screw. This is a device that traps pockets of water between a spiral screw that is rotated inside a sloping cylindrical sleeve. By turning the screw, the pockets of water taken from the tank or pond below were lifted to the height required. Without this mechanism, the hanging gardens would have been impossible to maintain, because the gardens were elevated and required a considerable amount of water from the River Euphrates flowing through the city.

The garden complex was square (approximately 150 m x 150 m), about 30m high, and was supported by sets of arched vaults, creating a number of ascending terraces. Each pillar and arched vault was made from baked brick and asphalt, and the pillars were hollowed out so that they could contain the roots of the largest trees. The actual floor of each terrace was constructed from stone beams about 5 m long, overlaid by reeds set in tar with two courses of baked bricks above the reeds, and sealed above by a layer of lead. This effectively prevented moisture filtering through to the floor below.

The screws conveying the water were operated manually, and although set alongside the stairs that took people to the uppermost roof terraces; the screws could not be seen. The terraces were arranged like a theatre so that the sunlight could reach every corner. It must have looked amazing to visitors, who would have marveled at the luxurious growth effectively hanging in the air. But the hanging gardens of Babylon were destined not to survive. Apparently there were several major earthquakes in the second century BC that destroyed the gardens. They had lasted about three hundred years.

The Exile in Babylon technically lasted seventy years, but when it was over only some of the Jews returned. In effect the 'Diaspora' or dispersion of the Jews has continued until the present day. The modern state of Israel was established in 1948 by the Allies after the Second World War[115], partly out of guilt and sympathy for the suffering of the Jews in the Holocaust. Many Jews who had retained their cultural and racial identity down the centuries then moved back to Israel. For them their exile was over. But there are still many people living in various countries around the world that can also trace their ancestral roots back to the original exiles from ancient Israel, and have managed to retain aspects of Jewish culture. Most of these people look to Jerusalem as their original home. A full return from Exile has therefore yet to be completed. But the Christian theologian, N T Wright, has a different interpretation of what God is doing in history. He strongly suggests that Jesus saw himself as embodying the real 'return from Exile' in his sacrificial death; see Wright (2012)[116]. In particular, the Passover meal that Jesus ate with his disciples the night before he died established the new covenant spoken of by Jeremiah; see Jeremiah 31:31. 'This (covenant) would be the establishment of the means by which 'sins would be forgiven'—in other words, the means by which God would deal with the sin that had caused Israel's exile and shame . . . . '[117] (p. 176). It follows that now, those who regard themselves in exile and those who don't, both have access to the Father through the Son; for the Exile is no longer an excuse. We may be seeing in our present day radical shifts in theological emphasis that have far reaching political repercussions.

References:

Wright, N T (2012) *Simply Jesus*. SPCK

Finkel, Irving (1988) 'The Hanging Gardens of Babylon,' In *The Seven Wonders of the Ancient World*, Edited by Peter Clayton and Martin Price, Routledge, New York, pp. 38 ff. ISBN 0-415-05036-7.

Resource questions

How radical is Jesus for you? How does he affect today's politics in the Middle East, particularly between Jew and Arab? Does the thought of Jesus' second coming make any impression on your daily life?

24

Water and birth

'. . . the time came for the baby to be born, and she gave birth to her firstborn, a son. She wrapped him in cloths and placed him in a manger, because there was no room for them in the inn.'

Luke 2:6, 7

'Water is the blood in our veins.!'
Levi Eshkol, Israeli Prime Minister, 1962 quoted by
Marq de Viliers in 'Water: The Fate of
Our Most Precious Resource',
First Mariner Books 2001

Read Psalm 139

Christmas is the celebration of Christ-mass, the time when we remember that Jesus was born as one of us, taking our human flesh, in order to conquer sin in that flesh. Our commemoration of Jesus' birth is however, carefully sanitized. Most pictures on Christmas cards or religious paintings show Jesus sitting benevolently on his mother Mary's knee, while she is clothed in flowing blue and white garments in an idealized and pristine stable. We are so familiar with the story of how Jesus was born that we forget the process of his birth. Birth is often painful for the mother, as it presumably was for Mary. The pain of childbirth was introduced into our creation by the sin of Adam and Eve in the Garden of Eden. God said that he would 'greatly increase your (Eve's) pains in childbearing; with pain you will give birth to children' (Genesis 3:16). Yet the pain and stress of the birth are quickly forgotten by the mother as she holds her new baby to her breast, full of joy that her baby is born, and bonding with the little one in a moment that she has spent nine months anticipating. Her view of the world at that moment is focused on her baby, on her joy that she has introduced another human being into the world. And water is at the heart of this occasion.

God's view of birth is not sanitized. This is made clear in Ezekiel's description of God's word to Jerusalem; see Ezekiel 16:4-7. Using the analogy of birth, Ezekiel describes how when Jerusalem was born, her cord was not cut; she was not washed with water, nor rubbed with salt or wrapped in cloths. No one had pity on her; she was thrown into the open field. God passed by and saw her 'kicking about in (her) blood' and said 'Live!' Subsequently, God made her grow such that she became 'the most beautiful of jewels'.

As we have repeatedly seen, water is crucial for life as we know it on this planet. From conception and birth to old age and death, our existence as human beings depends on water, not least because it is the primary component of the blood that courses through our veins and arteries to and from our hearts. The kidneys process 180 litres of water per day from the blood along with waste material dissolved in the urine or included in it as fine particulate matter[118]. However, as much as 80,000 litres per day are exchanged through the capillary walls of the cells in the body. Consequently, water molecules are continually passing through the body, spending no more than several hours in it, compared with, say, 9 days in a river, 6 months in a lake, 100 years in groundwater and more than a 1000 years in the oceans[119]. We wash our bodies with water, we swim in it, and we can drown in a depth of no more than 10 cm. Contrast that with the fact that for the first nine months of our existence following conception we were immersed in water in our mother's womb, where water filled our lungs. The umbilical cord linked us to our mother, through whom our developing body was resourced with critical chemicals at the appropriate times. Our lungs in the meantime were getting ready to take in air at our birth when the water was removed from them.

It was as true for Jesus, the Son of God, as it was for us. Having been conceived through the action of the Holy Spirit inside Mary's womb, his body and limbs developed in accordance with his DNA, as it were in secret, with only his heavenly Father aware of the details; see Psalm 139:13. Like all growing fetuses, the baby in Mary's womb was dependent on the umbilical cord which connected him with his mother until the moment of his birth in Bethlehem some 2000 years ago. Mary and Joseph had only just finished the long journey from their home in Nazareth, in the north of the country. Mary had probably been unable to do much walking because she was very pregnant with her first baby. If she had ridden on a donkey, she would have had to endure the up and down and side to side motions, as she wondered when they would arrive at their destination. On reaching Bethlehem they hoped to find a room in an inn where Mary could rest and have her baby. But other people had also made the journey to Bethlehem before them; they were all going to register with the Roman authorities for Caesar Augustus' census. When they arrived they found that 'there was no room for them in the inn' (Luke 2:7). This was Luke's conclusion. However, Bailey (2008) points out that on reaching Bethlehem they probably hoped to find a guest room with one of their relatives where Mary could rest and almost certainly have her baby. Therefore it is likely that they stayed in the crowded quarters of a family house above the shelter for the family's animals. Mary and Joseph made the best of less than salubrious conditions. They prepared themselves for the imminent birth of her baby. Mary would have experienced a very familiar sequence of events leading up to her baby's birth. It is probable that the membrane in her womb, containing the baby and the water in which it moved, ruptured, and the process of birth began. Mary would have experienced growing contractions in her lower body as her cervix dilated, and eventually the baby's head emerged, and Joseph or the local midwife gently took the baby as he came from his mother's body. When the baby was born he started to breathe air, and doubtless exercised his lungs in doing so. Joseph or a midwife wrapped the new born baby with soft, swaddling clothes that Mary had brought with her, and rested the baby on his mother's breast. The midwife had to wait until the umbilical cord stopped pulsating before tying and cutting it, and then delivering the placenta. The midwife tenderly and gently washed Mary and her child with water. The baby was sustained by milk from his mother's breasts. So, in various ways, water was an integral part of Jesus' birth as it was for us.

The long months of waiting for Mary and Joseph were over. Soon afterwards the lowly esteemed, rough and ready shepherds from the hills around Bethlehem arrived to see the baby. On the eighth day after his birth, Mary and Joseph had their son circumcised. After another five weeks they made the short journey from Bethlehem to the temple in Jerusalem to present Jesus to the LORD according to the law, and to sacrifice two pigeons. The elderly Simeon and Anna emerged from the shadows to prophesy over him. Later, the family received wise men who had journeyed far from a land in the east. Their visit alerted the Jewish authorities, and especially King Herod the Great, to the baby's birth. Herod was not able to rest until he thought he had annihilated the baby boy, whom he saw as a potential competitor for his throne. Jesus' birth led to the death of many young boys of his age in the district of Bethlehem, but Joseph was warned in a dream to take the family to Egypt. Jesus thus escaped Herod's assassination attempt, but the family were refugees for a few years. So the first years of Jesus' life were fraught with uncertainty, that is, until after Herod's death when the family was able to settle down safely in their new home town of Nazareth, some way from the political intrigues of Jerusalem.

Jesus was subject to the same physical laws and processes as we are. He grew up as a tender plant (Isaiah 53:2) before the LORD. His body functioned in the same way as ours in needing food and drink and disposing solid and liquid waste. He, too, was about 70% water, and as a carpenter and a friend of fishermen he would have been profoundly aware of the need for water in the environment, and the need to work to prepare food to eat. The day would come when he would be immersed in the waters of the River Jordan by his cousin John the Baptist. That event marked the transition between life in his family with his mother, brothers and sisters, and a ministry among his own people which eventually resulted in his death and resurrection. Water would be important then too. But first we continue to explore how water has important roles in promoting the good news of Jesus Christ.

Reference:

Bailey, K (2008) *Jesus through Middle Eastern eyes: Cultural studies in the gospels.* IVP Academic

Resource questions:

See if you can work out the water footprint of a baby

What does it mean for you that Jesus came into this world as a human baby?

25

Water and blood

Read Hebrews 9:11-28

Blood to the body is like water to the earth. Just as an irrigation system carries water, oxygen and nutrients along its channels to particular agricultural areas for surface plants, so blood circulating in arteries and veins carries water, oxygen and nutrients to the cells in the body. The body's circulatory system passes blood through the kidneys, which remove waste substances from the blood to form urine, which is subsequently excreted from the body. Similarly, much of the water available to vegetation through irrigation dissolves nutrients consisting of compounds of nitrogen and phosphorus and other chemicals necessary for plant growth. Surplus water drains through the soil and feeds groundwater aquifers and surface streams. Where an irrigation system provides water for agriculture there will usually be a drainage network designed to remove the excess (polluted) water.

Blood consists of a liquid called 'plasma', which is 90% water and in which are dissolved various other chemicals such as ions, hormones, and carbon dioxide[120]. The plasma carries red and white blood cells containing haemoglobin which facilitate the transport of oxygen to all body cells. Carbon dioxide is one of the basic waste products of the body and is removed primarily by being dissolved in the plasma. Our blood also carries water to the cells of our skin so that water discharged onto the skin's surface can evaporate as sweat. The process of evaporation, that is, turning water into vapour, requires energy, and in this case the energy is provided by excess heat generated by the body when raising its temperature to fight infection, or by the heat released when we exercise our limbs and muscles. In both cases heat is emitted from the body through the skin and through the air we breathe out. It is important that the body does not get too hot, and sweating tends to reduce its temperature to acceptable levels.

Just as humanity, and indeed all life, could not survive without the hydrological cycle driven by the radiation from the sun constantly renewing our supply of fresh water, so our bodies cannot survive without our hearts continuing to pump blood around our bodies. When the heart stops functioning, we die quickly because blood is no longer circulating and the brain in particular is starved of oxygen. All of us eventually reach the stage when our heart stops, whether naturally through old age or through illness or physical accident.

Theologically, blood is important because God stated that 'the life of a creature is in the blood'. He made this clear to Noah after the flood. God made a covenant with Noah, and gave him everything that lives and moves, that is, animals as well as plants, for food. The one restriction was that Noah was not to eat meat that still had the 'lifeblood' in it. There was a very good reason for this restriction as most diseases are carried by blood and transmitted by it through the body. Not eating blood was subsequently a factor in maintaining the health of the Israelites. Because of God's regard for blood, he demands from each man and animal an account of its lifeblood. If any man spills the blood of his fellow man, he will be held accountable by God because he has taken another man's life. In particular, whoever sheds the blood of man, by man his blood will be shed (Genesis 9:4-5). The salutary reason given for this is that man is made in God's image. In other words, man's blood is in some sense patterned on God's being. Blood is at the heart of God.

We do not have to read far into the Old Testament to realize how this truth is worked out. When the Israelites were planning to leave Egypt, God sent a final and terrible plague in which the first-born males of human beings and other animals died (Ex.11:4-5). The Israelites were given a way of escape from the plague. Each household was told to kill a lamb, according to strict instructions. Blood from the lamb was to be put onto the sides and tops of the doorframes of their houses (Ex.12:1-11). The lamb was eaten with unleavened bread, the last meal the Israelites had before they were rescued from Egypt. When the LORD went through Egypt to kill the first-born males, he saw the blood on the doorframes of the Israelite houses, and would not let the Angel of Death enter to kill the first-born sons of the Israelites (Ex.12:23). So the Angel 'passed over' the houses of those Israelites with lambs' blood on their doorframes. The Israelites acted in faith, trusting that the Angel of Death would do what God had promised. The blood, which represented the life of a lamb, protected and redeemed the life of the first-born. Subsequently, the first-born sons of the Israelites who did not die were consecrated to the LORD.

What happened that fateful night was remembered by subsequent generations of Israelites as they repeated the meal, called the Passover, each year. In Jesus' time, the Passover was the greatest of the Jewish festivals. It was during a Passover meal that Jesus shared with his disciples that he instituted what came to be known as the LORD's Supper, or Eucharist. During the meal Jesus made a new covenant with his disciples, and consequently with everyone who believe(s) in him, by inviting them to eat bread that he blessed, broke, and distributed, along with a cup of wine, which he also blessed and invited all to drink. Jesus said that the bread was his body and the wine his blood. In eating the bread and drinking the wine we feed on Jesus 'in our hearts, by faith, with thanksgiving'[121]. What is important for us here is the fact that Jesus referred to the wine as his blood. He was looking back to the rich sacrificial system of the Israelites, in which God's forgiveness of sins was coupled with animal sacrifices. Jesus saw himself as the Passover Lamb, who would be crucified outside Jerusalem, and whose blood would be shed, not to be put on doorposts, but to enable all people who put their trust in Jesus to be forgiven.

After the Israelites escaped from Egypt and the Egyptians, they journeyed through the wilderness on their way to the Promised Land. On the journey, the LORD made another covenant with the people. In sealing the covenant between the LORD and his people, Moses took the blood of bulls and goats, mixed it with water, and sprinkled it on the book of the law, on the people and on the altar on which he had sacrificed the animals; see Exodus 24:8. The writer to the Hebrews says that the sprinkled blood pointed to Jesus as the mediator of a future covenant; see Hebrews 12:24, for 'without the shedding of blood there is no forgiveness'; see Leviticus 17:11 and Hebrews 9:22.

When God made the covenant with his people he also gave them his laws and commandments, together with the rules for the sacrificial system to be carried out by the priests first in the tabernacle, and eventually in the temple. The blood of animal sacrifices became the focus of this system. The tabernacle was a tent which signified the LORD's presence among his people. It stood within an open court where the animals were sacrificed, and it had two sections, the Holy Place, and the Most Holy Place which the high priest would enter once a year on the Day of Atonement to sprinkle the blood of the animals sacrificed for the sins of all the people.

God pronounced his judgment on sinners: 'The person who sins is the one who will die'; see Ezekiel 18.4. Each person's sin resulted in a death sentence for the individual committing the sin. In his mercy, however, God decreed that an animal, a bull or goat, might be sacrificed to atone for sin, instead of requiring the death of the sinner. The sacrifice was made effective when the 'sinner' placed his hand on the head of the animal to be killed and so identified with it—acknowledging that the animal was dying in his place. The blood of the animal was then sprinkled around the base of the altar. It followed however, that if only past sins are forgiven on the basis of an animal sacrifice, further sacrifices are needed for future sins. The Jews came to realize that the blood of bulls and goats could not take away sins permanently, and that animal sacrifices are only a pattern of what God intends. But this raises the important question: What is the reality behind the pattern?

The writer to the Hebrews tells us that Jesus entered the 'real' tabernacle of heaven, taking his own blood to secure eternal redemption for us (Hebrews 9:12). His blood was shed as the Lamb of God, who the Apostle John states was crucified for the sins of the whole world (1 John 2:2). As a result, eternal life has been secured for us by the death and resurrection of the Lamb, 'slain from before the foundation of the world', because this was always God's intention. That life is available to us when we put our faith in him. We share in this life, not by having his blood sprinkled on us as in the Old Testament practices, but by participating in the foretaste of the heavenly banquet when we eat the flesh of the Son of Man and drink his blood; in other words, by regularly receiving the bread and wine in the Eucharist, which we do in memory of the Saviour who died for us, and who told us to do it. The wine represents the blood which sealed the new covenant, referred to by Jesus during the Last Supper with his disciples, and secured by his death on the cross. The New Testament writers are in no doubt that we receive atonement for our sins (Romans 3:25), are justified (Romans 5:9) and redeemed by his blood (Ephesians 1:7).

The writer to the Hebrews is even more explicit. He views Jesus as a faithful high priest who entered the real (heavenly) Most Holy Place to offer himself to God through the Spirit, taking his own blood to cleanse us from our sins (Hebrews 9:11). Peter says that 'we are redeemed by the precious blood of Christ, a lamb without spot or blemish' (1 Peter 1:19), while John says 'the blood of Jesus his Son purifies us from all sin' (1 John 1:7). So we know freedom from our sins by Jesus' blood (Revelation 1:5). He has purchased people for God from every tribe, language, people and nation; in other words, there is no group which will not have heard about what God has done in Christ (Revelation 9:5). In the new heaven and the new earth, there will be men and women who have overcome the evil one 'by the blood of the Lamb and by the word of their testimony' (Revelation 12:11). It is significant that the resurrected Jesus took his (raised) human body into the Father's presence, thereby taking the sacrifice he made on Calvary to the Father's throne. In so doing, Jesus united things in heaven and earth, restoring the relationship that was God's purpose for creation before the Fall.

We have come a long way from our assertion that blood is made up largely, but by no means exclusively, of water. The word 'blood' has crucial theological significance associated with it. We do well to ponder what the blood of Jesus has done and continues to do for us.

Resource questions:

What does the Eucharist or LORD's Supper mean for you?

When you think about drinking the wine as the blood of Jesus, what does this imply for your daily life?

26

Water into Wine

Read John 2:1-11

As I look back over my life I am aware that scarcely a year goes by
without attending one or more weddings; and that is not because I am
ordained! But it is still somewhat of a surprise that the Apostle John
should record that one of the first things Jesus did in his ministry was
to take his disciples along to a family wedding. (Actually John does
not say it was the marriage of a family member; but the presence of
Jesus' mother, with her immediate concern for the shortage of wine
at the feast could lead us to suspect that the bridegroom may have
been one of Jesus' brothers). During the course of the marriage feast,
things began to go badly wrong; a situation arose which would reflect
badly on the host family and on their provision for the feast. It seems
they had failed to anticipate the quantity of wine actually needed for
the wedding; maybe they hadn't taken into account the large number

of disciples accompanying Jesus! This sort of thing would be the talking point of the community for months if not years to come. Jesus' mother, Mary, noticed that something was wrong and simply stated the problem to Jesus. From his seemingly abrupt reply it appears that she was asking her son to do something which would draw on his innate resources, before it was appropriate. He responded that it was not yet time for him to fulfill the purpose for which he had been born. But that was enough for Mary; she instructed the servants to do whatever Jesus told them to do. Then she disappeared from the story John was telling. It is as though John was presenting Jesus as the one who was taking over the story line from her, and what mattered now was the relationship between Jesus and his heavenly Father.

Jesus instructed the servants to fill six large stone pots to the brim with water. These were large pots, each holding many litres, and were intended for purification rites. There are subtle hints here by John that facilities used to meet the requirements of the Mosaic Law were being appropriated to provide services for the new covenant that Jesus had come to inaugurate. In addition, the number six is an imperfect number indicating that what Jesus was about to do was still going to be incomplete. It must have taken the servants a number of journeys with their portable jars or skins to and from the local spring of fresh water to complete the filling of the purification pots. Once these were full, Jesus told them to draw out some of the water into a cup or goblet and to take it to the master of the feast. We can imagine that the servants did so obediently, yet somewhat apprehensively, for they had put the water into the jars, and they had no reason to think that it was anything other than water. To their surprise, the master exclaimed that what he was drinking was the best wine of the feast. Why had the bridegroom kept it until now? Surely, the most appropriate strategy was to start with the best wine when everybody's palate was fresh, and then to resort to the poorer wine when people were less aware of the distinction between the good and the less good? The master did not know where this excellent wine had come from, and called the bridegroom to account for it. But the bridegroom was also unaware of its source, even though the servants knew.

Such misunderstanding is typical of the way in which John highlights the faltering insight of people into who Jesus was and what he was about. After all, for John, Jesus was the real bridegroom.

What Jesus had done to the water was an 'epiphany' for his disciples and John's readers (that is, a manifestation or appearing) of who Jesus was, a sign which pointed to him as the promised Messiah who gives eternal life to those who believe in him. In other words, Jesus' actions at the wedding in Cana in Galilee pointed to his glory: it was a sign of something supremely profound about him in meeting our needs. The vast quantity of mature wine, which seems to us a grossly over-extravagant gesture, pointed forward to the extravagant and overwhelmingly generous action of Jesus when he died on a Roman cross for our sins, and to the sufficiency of his blood to cleanse us from all unrighteousness. The sign pointed to who Jesus was, and it was enough to persuade Jesus' disciples to put their faith in him. Their faith was still relatively uninformed, but they had crossed the starting line of their discipleship and there was no going back. Note that Jesus did not create the wine in the pots without first having water poured into them, just as he did not feed the five thousand people without having five small loaves of bread and two small fishes in his hands. Jesus involves us in his actions, taking what we bring, blessing and multiplying it.

The superabundance of wine was generated out of water, just as wine was produced from rainfall on the vineyards of Israel. The wine came from grapes that were the fruit of vines well supplied with water. In Isaiah 5:1-7, Israel is likened to a vineyard that a man planted on a fertile hillside. He provided everything that was needed for the vines to be fruitful. He built a watch tower and a winepress to extract the juice from the good grapes he expected. But the vines yielded only bad fruit, and so the investment was wasted. What was the owner to do? He would have to remove what he had built, and let it revert to being open land again allowing briars and thorns to grow there. Then we learn that God was the owner of the vineyard, and the houses of Israel and Judah were the 'garden of his delight'. God looked for justice and righteousness in his people's dealings with him and each other, but only saw bloodshed and heard cries of distress. In other words, his people failed to live according to his requirements and expectations. As a consequence God would no longer protect his vineyard. He would remove the walls that surrounded it and let it become an arid wasteland with little rainfall; that is, God would bring judgment on his people.

Water into wine

John's account of the wedding at Cana in Galilee is full of allegory. John says that the miracle took place on the 'third day', but we are not told the reference time of this third day. He is almost certainly referring to the resurrection and exaltation of Jesus. Similarly, the wedding feast is a foretaste of the 'marriage supper of the Lamb'. The empty water jugs reflect the emptiness of the Jewish purification ceremonies; and they are now filled with the new wine of the last times. Also, wine is the symbol of blood. When Moses came before Pharaoh with the message from God: 'Let my people go', the first miracle and sign that he did was turning water into blood. John makes much of the life-giving nature of the water and wine that Jesus will give to his disciples to drink. In particular, when Jesus hung dead on the cross, the centurion pierced his side with his spear, and out flowed blood and water, symbolizing the Christian sacraments of the Eucharist and Baptism.

Today, churches, especially Nestorian, Coptic, Abyssinian, Chaldean and Orthodox in the Middle East, drink wine ritually at weddings in memory of the wedding at Cana[122]. Lady Drower (1956) records that Mandaean wine is made from water collected by a priest in a bowl from a running source. Fresh grapes or raisins and dates, representing the vine of female fertility and the date-palm of male fertility, are put into the bowl, and the priest presses the fruit with his fingers until the water is obviously coloured. The bride drinks this 'wine' three times and the bridegroom seven times.

We have the advantage of looking back to Jesus Christ as the One in whom so many of the Old Testament promises were fulfilled. When we fail him, we have forgiveness if we truly repent. But we need to maintain a right relationship with God in Christ. Just as Jesus was able to transform water into the best of wines, so he is able to transform us to be men and women who richly benefit others. He can change us into something that like salt and light benefits the whole community in its life. We are ordinary people, but Jesus takes us as we are and transforms us into something extra-ordinary. Thanks be to God who has given us this privilege and opportunity.

Reference

Drower, E S (1956) *Water into Wine: A study of ritual idiom in the Middle East*. John Murray, London

Resource questions:

How do you maintain a right relationship with God?

What does it mean for you to be 'pruned' by your heavenly Father?

Think when you last took a risk for Jesus? What happened?

27

Living water

'A generous man will prosper,
he who refreshes others will be himself refreshed.'
Proverbs 11:25

Read John 4:1-8

The Alps in Switzerland receive large amounts of snow each winter—'snow on snow'. In the springtime, as the temperature rises the snow may rapidly melt, and the melt water cascades down the steep mountain slopes producing in places dramatic waterfalls that generate clouds of mist and spray. Walking in the mountains at his time carries its own dangers. What is so impressive are the streams of water noisily making for lower ground bringing life and refreshment for lakes and fields far below. The water is coloured a steely grey-green, having percolated through alpine vegetation and crashed over rocks and gravel. The turbulent nature of the flow accelerates the assimilation of oxygen into the water counteracting any tendency of the percolation to limit the life-giving properties of the water for the ecosystems downstream by picking up any harmful substances. The water is effectively alive as it brings both the threat of flooding and the promise of inherent resources to promote growth and fruitfulness in the lowlands.

As a child in the 1940s in the west of England the Alps were a distant dream. I was aware that my grandparents had piped water into their homes, yet the supply was intermittent. Occasionally I was dispatched with a shiny steel can to collect water from a spring in the yard behind the local hotel in Hay-on-Wye, just down the road. Unlike the piped water supply, which originated from a small reservoir fed by streams from the hills above my grandfather's fields, the spring was constant in its flow. Its source was an underground aquifer which acted like an extensive inaccessible reservoir, receiving an irregular but sufficient supply of seasonable water from rainfall on the hills above the town. But though the supply was irregular, the size of the underground storage ironed out the variations in the inflow so that the flow I observed emerging from the spring was almost constant.

A spring by its very nature is dynamic. Water continually flows from it out onto the ground surface, forming a brook or stream. Moving water absorbs oxygen more easily and is therefore likely to be relatively free of bacteria, and is safer to drink than water that has been standing for a while. I was made very aware of this fact when I visited Sekamuli, a village near a town called Luweero, an hour and a half's drive from Kampala in Uganda[123], and very different from my grandparents' home town. With very few streams in the area and some swamps containing polluted water that evaporates in the dry season, the local population is dependent on the few and widely distributed boreholes to provide fresh water; see Chapter 8 of this book. Fortunately, the boreholes are reliable, provided they are deep enough to reach the underlying aquifer. Through (artificial) wells with hand pumps, the population has regular access to fresh water which they can drink safely after boiling. (Indeed, the water in some of the wells can be drunk without being boiled, but because this is not true for all wells the general advice is to boil the water before drinking so that any bacteria can be killed before the water is consumed.)

Where springs come out onto the ground surface from the underground aquifers, they tend to be the focus of water supply for the local community. It is important to make use of the emerging water in a way that gives maximum benefit to the community. For example, water collected immediately from the spring as it emerges from the ground is the safest to use for human consumption. Immediately downstream of where the water is collected for drinking purposes, it can be used for food preparation, then for washing and only lastly for animals to use. Any residual water can be used to sustain the wetlands through the dry season.

Before the development of piped networks to supply fresh water from a sizeable resource people were dependent for their water on springs and wells. For many centuries, springs where water emerges directly from the ground, have attracted attention, not only because many provided safe water for drinking, but because they became associated with magic, mystery and religious practice. This was particularly true in Medieval England. As Ian Bradley points out in his book: 'Water—A Spiritual, History', springs were considered to be sacred. They were reputed to have curative powers, and their water was regularly used for Christian baptisms. Villagers would 'dress' their wells in the Spring: they were a focal point for village life.

Aside from their spiritual significance, springs are especially welcome in developing countries, because they are often a more reliable source of clean, fresh water than surface streams. As we have stated above, water from springs tends to be relatively uncontaminated by bacteria. To many, springs are a mystery. We cannot see the groundwater aquifer, and therefore fail to appreciate the pressures that drive the water from the aquifer through the spring outlet. In his conversations with Nicodemus (John 3) and the woman at the well (John 4), Jesus uses water as an analogy to explain the way in which the Holy Spirit can be regarded as a bubbling resource from within a person that benefits others. The image of 'living' water flowing from the heart of the Christian can remind us of the rivers flowing from Eden to water the earth, or the river, spoken of by Ezekiel and by the Apostle John in Revelation, flowing from the throne of God. (We will consider this river more closely in the last chapter of this book). What this kind of image points to is the way in which each Christian can be a resource for others. Such a picture is reminiscent of the water fountains we see in parks, gardens and public squares. Some fountains are provided especially for people to drink; others discharge water under pressure into the air, often with a gentle twinkling sound as the water leaves the nozzle or pipe. The pressure driving the water may be generated by storing the water at some height above the nozzle and letting it flow down the pipe to the outlet. Such fountains make the surrounding environment more humid, and thereby convey a general sense of wellbeing to those around.

Fountains and water jets are some of the basic components of what are called 'water' parks. The jeux d'eau ('watergames') at Schloss Hellbrunn in Salzburg, built by Markus Sittikus von Hohenems, Prince-Archbishop of Salzburg to amuse his guests, create a very good example of a park where practical jokes may be played on guests or visitors[124]. One of the first jokes consists of stone seats around a stone dining table in the park adjacent to the Palace. A water conduit conveys water to the seats of the guests. When a particular mechanism is activated, fountains spray water into the seats, making the guests very wet. Then throughout the gardens there are a number of hidden fountains that surprise and spray guests as they move around. It is well worth a visit, but be prepared to get wet!

Whether the stream of water coming from a fountain is large or small, whether the flow is laminar or turbulent, whether it is hot or cold, the analogy with the Spirit being a resource for human hearts means that there are adequate spiritual resources in the Christian community for all in need.

Reference:

Bradley, I (2012) *Water: A Spiritual History.* Bloomsbury

Resource questions:

What water parks and gardens have you visited? What has impressed you most about them?

Living water is fresh water. How do we keep the water of the Spirit fresh in us?

28

Into deep water

'Only a fool tests the depth of the water with both feet'
African proverb

Read Luke 5:1-11

Many of us have learned how to swim. It is regarded as a most important skill, especially in countries such as the Netherlands, which abounds in canals and stretches of open water. Dutch children are encouraged to learn how to swim from an early age, taking tests to earn their coveted swimming badges. At first, when we learn to swim we have buoyancy aids to reassure us that we can float on the surface of the water. We generally grow in confidence as we practice moving our arms and hands to increase the lift on our bodies when in the water. As we become more confident, we are able to dispense with the aids, provided we know we can touch the bottom of the swimming pool. But what about swimming in the deep end of the pool, where the more advanced swimmers dive in? We know we cannot put a foot on the bottom of the pool there and still have our noses or mouths in the air. For most people it is simply a matter of confidence. Can we be assured that we can easily remain afloat even when the strokes with our arms are restricted? Once we have such confidence, we can swim more easily, and provided we develop our physical strength, we can take on being further 'out of our depth'. This involves taking risks, that is, putting ourselves in the position where on a very few rare occasions we might fail to achieve our objective. For example, when crossing a road busy with traffic we would normally choose to cross at a pedestrian crossing, where the traffic is required to stop due to red lights, or on a road when there is only the remotest possibility that a vehicle would hit us and knock us down. Of course there is still the need for caution. When we are swimming in the sea we should be aware of potentially dangerous conditions when we could be challenged by unusual circumstances, such as a rip current or an undertow, both of which might carry us out to sea. To avoid the rip current we need to swim across its flow direction rather than against it, as soon as we become aware of what is happening. Similarly, with the undertow we need to get to the surface of the water where the flow is less strong. It is important to develop an understanding of how water behaves in our environment if we are to enjoy its benefits.

One day Jesus was teaching beside the Sea of Galilee. The crowd was thronging around him, and they began to press him back towards the water. He needed more space, so he asked a fisherman to push his boat out a little from the shore so that he, Jesus, could stand in it and teach the crowd. It proved to be an ideal solution to the problem. Once he had finished teaching, Jesus told the fisherman (whose name was Simon, and whom Jesus was later to call Peter) to launch out into the deep and let down his nets for a catch of fish. This was not what Simon expected him to say. Simon had been fishing all night and had caught nothing; yet everyone knew that night-time was the best time to fish, not the day. During the night, the larger fish would come into the shallower waters near the shore to feed on the small minnows which abounded in water that had been warmed by the sun during the day. By day, the larger fish would be long gone, so it made no sense to go fishing at that time. But Simon did not argue. He agreed to do what Jesus said and set out for the deeper water. Perhaps reluctantly he and his colleagues let down the net, only to be taken by surprise: the nets were being stretched. They were becoming taut. There was obviously a huge shoal of fish down there and the nets were full. They were even going to break. They called on a second boat to join them in order to raise the nets before the boat sank. They barely made it in time; see Luke 5:1-11.

Simon realised that he had misjudged this wandering rabbi. He had not really believed what Jesus said was true. He had gone through the motions of doing what Jesus said, but he was sure before he started that there were no fish there. So when he was proved wrong and Jesus' words were proved right, Simon could only think about his inadequacy and contradiction in the presence of Jesus. He fell to his knees in front of Jesus and said to him, 'Depart from me, for I am a sinful man'. He had completely lost confidence in his own ability. But Jesus did not want to demoralise Simon; rather he wanted to challenge and encourage him to believe and act on what he told him.

By demonstrating that his words carried more weight than Simon's experience, Simon could trust him, whatever Jesus told him to do in the future. In Mark 1:17 Jesus told the disciples that he would make them fishers of men. His words were conclusive for the disciples. They left everything to become the people Jesus wanted them to be, to follow him in order to 'catch' people for the kingdom of God.

Staggered by the haul of fish, Simon had been jolted into recognizing Jesus as his LORD. There were few questions left in the mind of this simple fisherman. It may seem strange that reports of the greater miracles of healing and expelling demons had not moved Simon earlier. Sometimes it is the simplest things that bring a person to realize that Jesus Christ is LORD. But by whatever means, it is vital that each of us realizes, as Simon did, who Jesus is. Acknowledging Jesus as LORD raised in Simon's mind a vivid picture of himself as 'a sinful man'. Suppose I compared myself with other people that I select; I might have reason to regard myself as being better in some sense (not that I would want to do so!). But when I compare myself to Jesus, His perfect purity is in sharp contrast to my sinfulness, of which I become very aware. Maybe Simon, knowing himself, confessed his own lack of truth and goodness, and his need. This confession of sin and failure in our lives is hard to admit—until we have done so. Then we discover that, once we have acknowledged that we fall short of God's standards, we are set free. For once we see ourselves as sinful we become more ready to hear the healing words of Jesus. For he said to Peter, 'Don't be afraid; from now on you will catch men' (v. 10). As those who sin, we are not afraid of God, for we know He forgives us, and transforms us as well. The essential nature of what Jesus told Peter is that from now on life would be different! Peter and his two companions left everything there on the beach—including the huge haul of fish—and followed Jesus. Their skills and experiences as commercial fishermen were willingly set aside. As disciples or followers of Jesus Christ, the whole of their lives were being radically changed. They had passed from independence of God to dependence on Jesus. But they were still only beginners in the kingdom of God, and they had much to learn. Within three years they would be empowered by Jesus through the Holy Spirit to convince men of the good news of Jesus Christ, confirming the nature of their life by signs and miracles.

After they experienced his power and grace, Jesus immediately called the disciples to follow him. Jesus does not call everybody in this way. Sometimes people can experience his power, but Jesus does not call them to follow him immediately. However, once someone hears his call, that person needs to weigh up the cost of pursuing the calling while not delaying unnecessarily in deciding to follow. Jesus is more likely to expect us to respond immediately than to delay while agonizing over what we should do. But for each person the call is different and personal. What Jesus does is call us to be prepared to take risks for him; that is, to be prepared to trust God by going outside our comfort zone. In so doing we will experience that God is trustworthy in every circumstance. God give us grace to respond in faith to Jesus Christ as LORD and Saviour.

Resource questions:

What professional skills do you have? How could these skills be adapted for the Kingdom of God?

What is God calling you to do with your life?

29

Walking on water

'Mack walked to the edge of the dock and looked down. The water lapped only about a foot below where he stood, but it might as well have been a hundred feet. . . . He looked back at Jesus, who was still chuckling.

'Peter had the same problem: How to get out of the boat. It's just like stepping off a one-foot-high stair. Nothing to it.'

'Will my feet get wet?' queried Mack.

'Of course, water is still wet.'

Again Mack looked down at the water and back at Jesus. 'Then why is this so hard for me?"

William P. Young, 'The Shack', Windblown Media

Read Matthew 14:22-36

It had been a busy day. Jesus and his disciples had been surrounded by the crowd of people who followed Jesus wherever he went. He had talked with them and taught them. He had healed many from different diseases and disabilities. Jesus needed some quality time with his heavenly Father. So when evening came, he sent the disciples off to the other side of the Sea of Galilee, implying that he would catch up with them later. The disciples got into a boat that was big enough for them all, and started rowing the 10 km or so to the other side. They took it in turns to row; the crossing would take them a few hours. At first they made reasonable progress, that is, until a storm blew up.

The Sea of Galilee is in the Jordan rift valley and is vulnerable to the sudden generation of storms characterized by high winds. These made rowing difficult. The wind had got up and the disciples were making little progress. It was the middle watch of the night, which meant that it was about three o'clock in the morning and just beginning to get light. They longed to be at the shore, but they were still some way off.

Then they saw a figure walking towards them—on the water. Their immediate thought was that it was a ghost, and they cried out in fear. It took Peter a little time to realize that there was something very familiar about the figure. It was Jesus. But he looked as if he was going to walk on past them. Peter was not one to let that happen. He called out to Jesus. He was so excited that he wanted to get out of the boat and go to Jesus—if Jesus could walk on water, he could too. But Peter realized, instinctively perhaps, that he needed Jesus' authoritative permission to do so, so he asked Jesus to call him over. Jesus did so and Peter got out of the boat. He put his foot firmly onto the water while sitting on the side of the boat. Then the other foot, and he pushed off and started walking towards Jesus—on the water. He was doing the impossible! But as he walked, he suddenly realized what he was doing, and how inconsistent it was with all of his previous experience of water. He had been looking at Jesus; now he took his eyes away from him and looked down at his own feet. They were disappearing under the waves as he began to sink. Peter appealed to Jesus to save him from drowning. Immediately, Jesus was at hand, pulling him up and helping him into the boat. The wind and waves stopped as soon as they were in the boat. The disciples were amazed at what they had witnessed. They did not know what to think. All that they had learned by experience about water and storms had been contradicted. People could walk on water, and storms could be stilled.

So how did Jesus walk on water—or Peter for that matter? There have been some fanciful explanations made, such as Jesus was walking on a sandbank or a shoal of large fish, or on ice or close to the shore. It is unlikely that Jesus had a way of strengthening the surface tension of the water, as there were waves on the surface, though he could have altered the buoyancy effect of having his feet somewhat in the water as he walked. Leonardo da Vinci sketched some floats which could be tied to the feet which would keep a person upright and on the surface of the water. (Indeed, Leonardo Da Vinci had a life-long fascination with water[125].) But it is unlikely that Jesus would have used (or even needed) such a solution, otherwise it would have been made clear in the gospel accounts. The precise explanation will have to remain a mystery for now. Anyway, as creator, Jesus knew how to take advantage of the physical laws and processes in the universe.

But what was the real point of the story? Jesus had spent a long time praying that night, and he was aware of the difficulties that his disciples were in. Then he chose to come and walk by them on the water, when they were getting somewhat desperate at their lack of progress. Jesus was walking effortlessly, whereas they were labouring at their oars. Jesus went as if to pass them by, waiting for them to call for his help. Fortunately Peter did so, and in stepping out in faith he learned more about himself and the power o his master.

What can we learn from this story about walking on water? We have to remember that it was not only Jesus who walked on water, so also did Peter, a sinful man like the rest of us. We can deduce from this that Jesus did not walk on water because of his divine nature— Peter was certainly not divine, and yet he walked on water too! This miracle came about through the Holy Spirit with them and in them. In his best-selling novel 'The Shack', William Young describes how Jesus and the hero of the book walk on water in order to cross a lake[126]. It seems incredible and yet strangely possible. It is the stuff of dreams, such as many people have of flying. Some dream that they are able to slide along the path they are walking along, finding themselves continuing to float a few feet off the ground, upright, and able to change direction in the air. Such dreams are not uncommon. Some would say that this is an indication that we are meant to fly or walk on water in the new heaven and the new earth. Certainly in this creation, to do either would have to be by faith; for Jesus, although divine, did not walk on the water in Lake Galilee out of his divinity but through the Holy Spirit with him and in him. Elsewhere, he talks about faith as a grain of mustard seed being sufficient to move a mountain; such faith would surely be needed to walk on water.

There is no record in the New Testament that Jesus or Peter walked on water again, and Peter does not reflect in later life in his epistles on this experience, remarkable as it must have been. In other words, walking on water was a one-off event. Yet the history of the early Christian church is replete with signs and wonders, and there are apocryphal stories of Christians walking on water, as do some devotees of other religions. Yet here we stray into fantastic and unconfirmed accounts, and we do well to contain ourselves with the biblical record.

Walking on water

What appears to be evident is that our scientific knowledge has done much to prevent us expecting God to interact miraculously with our physical world. This has led many Christians to become unsure of their faith or to minimise their expectations of God to act purposefully in their lives. We need to recover our courage and to acknowledge that the kingdom of God is among us; that is, God is present in our lives, and he works miracles in and through us; see C S Lewis' defence of miracles in his book 'Miracles' (1947).

To take part in the miraculous, that is, in something that is more than just coincidence but difficult to explain, needs faith, courage, expectation and hope. To do the unexpected, or something outside the normal scheme of things, which appears to contradict the physical rules of the situation, requires bravery, confidence, trust and commitment. God give us grace to embrace the miraculous when it comes, especially when we least expect it.

References

Young, W P (2007) *'The Shack'*. Windblown Media

Lewis, C S (1947) *'Miracles'*. Collins, Fontana Books

Ortberg, J (2001) *If you want to walk on water—you have to get out of the boat!!* Zondervan

Resource questions:

Why should we want to walk on water?

What would walking on water do for your faith?

30

Calming the storm

'Whenever I am troubled and lost in deep despair,
I bundle all my troubles up and go to God in prayer.
I tell him I am heartsick, and lost and lonely too,
And my mind is deeply burdened, and I don't know what
to do.
But I know he stilled the tempest and calmed the angry sea,
And I humbly ask that in his grace he'll do the same for me.
And then I just keep quiet, thinking only thoughts of
peace,
And abiding there in stillness my restless murmurings
cease.'

Written by a patient at Fairmile
Psychiatric Hospital, Berks, UK, circa 1982
© Roland K Price

Read Luke 8:22-27; Matthew 8:23-27

When writing this particular chapter, the media were full of super-storm Sandy which had just made landfall close to New York[127]. Huge damage had been done, with subways flooded, homes and buildings destroyed through wind damage and fire, and many people were killed. Sandy was termed a super-storm because of its large spatial extent rather than its central intensity as in a hurricane. Like a hurricane, the super-storm gained its energy from the warm ocean with large amounts of water vapour evaporated from the ocean's surface. These storms develop into intense disturbances in the atmosphere, characterized by strong winds, low pressures and possibly heavy precipitation. Generally, in the northern hemisphere, the rotation of the earth ensures that the winds in a hurricane or super-storm blow in an anti-clockwise direction.

The interaction between the atmosphere and the ocean is very complex. The difference in temperature between the ocean and the air above it can lead to a considerable exchange of energy as warm water vapour evaporates from the ocean and rises in the atmosphere. The growing energy of the water vapour locally leads to a pronounced decrease in the air pressure with associated anti-clockwise rotation of winds about the storm centre (at least in the northern hemisphere). In turn, the wind exchanges some of its considerable energy with the water by dragging the water surface and creating waves. These waves can travel long distances away from the centre of the storm: they may generate heavy swell with sizeable breakers on a beach where there is no sign of the storm that was responsible for them. In effect, they are messengers conveying information about the storm. They are called 'short' waves, because the distance between successive wave peaks is usually considerably less than the depth of the ocean.

Another wave form is called a 'long' wave. This type of wave has a length considerably greater than the depth of the ocean. It travels much faster than short waves: as fast as a high speed train in fact. Long waves are generated by the storm in two ways. The first is by the wind forcing the water in the direction it is blowing. If the wind is blowing towards the coast, this will act as a barrier or wall causing the water level to rise. The rotation of the earth then diverts the water that is piled up against the coast to move along the coastline in what is called a 'surge'. The second way in which a long wave is generated by the storm is through the (extreme) low pressure at the storm centre. This causes the ocean surface to be raised, resulting in a separate component of the surge generated by the storm. The surge travels along the coast in a similar way to tides in the area.

It is important to distinguish between the disturbance and the water, because the surge may travel hundreds of kilometers, whereas the water may barely move at all. The same is true of the ocean waves that travel across huge distances from the generating storm before making land fall on a distant continent. The water contributes to the waves by particles turning locally in approximately vertical circles.

A serious consequence of storms over the North Sea is the surges they generate in the ocean along the Scottish, English and Dutch coasts. The wind of a storm over the middle of the North Sea circulates around the storm centre in an anti-clockwise direction. It piles up water against the eastern Scottish coast. This water then moves southwards as a surge, driven by the Coriolis force due to the rotation of the earth. It moves down the east coast of England into the funnel formed by the narrowing channel between England and Belgium and France. Here the height of the surge increases and then the surge travels back up the eastern side of the North Sea along the Dutch coast. An extreme surge devastated coastal areas along East Anglia in England and along the Zeeland Dutch coast in 1953[128]. The event was so serious it prompted the construction of the world famous Oesterschldt barriers that now protect Zeeland, and the Thames barrier that is raised when central London is threatened.

Lake Galilee is too small for any surges or currents to be noticeably affected by the Coriolis force. The lake is situated in the Jordan rift valley and has the upper catchment of the Jordan River draining into it; the river continues from the lake to the Dead Sea. The nature of the rift valley means that the lake surface, which is approximately 13km by 21km, is about 220 m below sea level and is surrounded by hills. Strong winds can be generated over the lake due to the larger circulation of the atmosphere across the rift valley. These winds can arise in a very short period of time, and the storms can be violent. Jesus' disciples, who were fishermen by profession, were well aware of such storms.

Boats were an important means of transport between different towns and villages along the coast of the lake. Rather than take the longer route around the coast, people regularly took the more direct route across the lake. So it was that on one occasion, having spent the day teaching and healing in one place, Jesus decided to go across the lake to the other side. The disciples provided the boat and their expertise in sailing it. Having instructed them where to go, Jesus lay down in the prow of the boat to sleep. He was obviously tired after a busy day, and he let the disciples manage the task of getting to the other side of the lake. It was evening, and all expected a peaceful crossing. But this was not to be. A storm suddenly got up, and it was more severe than usual. Despite some of the disciples being experienced sailors, they had rarely seen conditions deteriorate to the extent that waves were being whipped up by the wind and beginning to break over the side of the boat. What is more, the boat began to fill with water faster than they could bail it out. The situation was going from bad to worse; and yet Jesus continued sleeping. The disciples were becoming not just concerned but desperate with the thought of what might happen, especially as their Master remained asleep. Jesus had told them they were going to the other side of the lake;

Stilling the storm

that should have been sufficient for them to know that they would arrive safely, and that their heavenly Father would ensure their safety. But the disciples had taken their minds off the destination and were focused on the immediate situation, which looked dire. What were they to do? They decided there was only one thing for it: they had to wake Jesus up and make him aware of the situation. After all, his life was at stake as well as theirs. So they woke him urgently. Once awake, Jesus took immediate action; he spoke to the inanimate storm. In so doing, his creative word controlled the wind so that the huge amount of energy involved was dissipated so rapidly that it appeared instantaneous. The wind disappeared, and the waves calmed down as if the water was glycerin. One moment the disciples were facing the frightening prospect of being drowned; the next moment everything was calm—all because their master spoke to the situation. They were overcome with relief and amazement. Who was this man they followed? How did even the wind and waves obey him? They were experienced fishermen; they had been in similar situations before and they knew the impossibility of what had just happened. Jesus said to them 'O you of little faith!' They should have trusted that Jesus would have had everything under control. He would not have allowed then to drown, especially when he was with them. The fact was, however, they had not had the faith to believe that Jesus would save them, and certainly not that he could control the wind and waves. They were now aware. Jesus was far more than a challenging teacher and amazing healer: he even had a powerful influence over huge forces of nature, and in particular the weather.

With our theological perspective of Jesus as the divine Son of God who was present with his heavenly Father at creation, and who as the Word of God was instrumental with the Father in facilitating creation out of the waters of chaos, we can read this account of Jesus stilling the storm and see the connection between him in his earthly ministry and him as creator with the Father. We lose some of the impact that the experience must have had on the disciples. They thought they knew Jesus and who he was, but they were continually being brought up short as they realized he was far more than they could reasonably appreciate or even imagine. Time and again, they had to change their minds about him as he made known to them more aspects of his being. How could such awesome behaviour be focused on one human being?

And where did the enormous energy that was transferred from the wind to the waves go in an instant? How did Jesus manage to calm both wind and waves simultaneously? This account brings to life the amazing control that God is described as having in the book of Job (see Job 38). But at least one question remains—can we have the power to do what Jesus did to the storm and the waves, when really needed of course? Elijah prayed in faith that it would not rain for three and a half years—and then that it would (see James 5:17, 18), and it did. Where there is a need to control or alter the weather there is the possibility that followers of Jesus can do what he did or even greater things (see John 14:12b), provided they too pray with faith that God will do what they request, and that their lives are lived out in obedience to his will and purpose. However, God's will and purpose is sovereign. We consider in Chapter 35 the occasion when the Apostle Paul was shipwrecked on his way to Rome. God had a purpose in the shipwreck, namely, to confirm to Paul that he would bring him safely through all the dangerous circumstances of his life, and to demonstrate to his fellow travelers, including hardened sailors and cynical soldiers, that God is interested in them.

Resource questions:

To what extent can we control floods and the weather?

How can we couple our experiences of nature with God's involvement?

How important is social justice in respect to the distribution of water, and to excess and deficiency of water?

'Take heart; do not be afraid.' How does Jesus help calm your fears?

31

Water and healing

*'I understood when I was just a child that without water,
everything dies. I didn't understand until much later
that no one 'owns' water. It might rise on your property,
but it just passes through. You can use it, and abuse
it, but it is not yours to own. It is part of the global
commons, not 'property' but part of our life support
system.'*
Marq de Villiers, 'Water – The Fate of
Our Most Precious Resource',
First Mariner Books, 2001

*'God, Who for the salvation of the human race has built
your greatest mysteries upon this substance (water), . . .
May this your creation be a vessel of divine grace to
dispel demons and sicknesses, so that everything that it
is sprinkled on in the homes and buildings of the faithful
will be rid of all unclean and harmful things. . . . Amen.'*
Prayer for the blessing of water, taken from the
Catholic liturgy for the Feast of the Epiphany

Read Psalm 139

Although the shape and characteristic features of the human body
persist in a person as they grow older, the physical material that gives
the body substance is always changing (Ps 139:1-4, 13-16). In a way, the
body is like a river with water molecules flowing through it; the river
channel has a seemingly permanent form and shape, yet the water in
it is continually changing. It is humbling to think that we share water
molecules with all living things; for example, water is extracted from
a river for water supply at one point then is returned to the river after
use (and hopefully, following the removal of pollutants), only to be

used by another community downstream. So water molecules are in transit through us to other people or other forms of life. The river that you see today will never be the same again. Our bodies are like that. The transient nature of our physical bodies belies the permanence of what it is to be us. You are uniquely you, even though the atoms and molecules that make up your body are continually changing. You have a brand new stomach lining every 4 days, and your skin is renewed every 30 days[129]. Your liver is renewed every 6 weeks, and even your skeleton is replaced every 3 months. Your blood plasma renews itself every 10 days, and white blood cells renew themselves every 2 to 3 weeks. On the other hand, your red blood cells typically take 120 days to renew themselves. Your entire blood resource is renewed every 3 to 4 months. About 98% of your body is renewed on average every seven years for a woman and eight years for a man.

Obviously, I am much more than the sum of my body parts. There is also a set of information patterns or templates associated with my DNA that hold the design of my body, and still more, a huge store of bits of information that record my experiences and which can be reconstructed in the form of memories. Then there is the whole issue of consciousness and what enables us to be aware of ourselves, to distinguish ourselves from others, and to relate to others as separate entities. We are extremely complex beings. The mental capacity of each human individual in terms of numbers of bits of information that can be stored exceeds the number of stars in the universe. This capacity is retained despite the molecules in our bodies changing completely over time. We truly are amazing creatures of an amazing God!

Biologists studying the human body at the cellular level describe how water interacts with each cell to give it substance and structure, and to transport materials into and out of the cell. There are sophisticated interactions between the water molecules and the different component elements of a cell. Because water is such an integral part of each cell, the healing propensity of each cell and organ of the body is strongly dependent on water, whether correcting the balance of constituents in a cell or transporting critical substances around the body. The fact is that the body contains a number of separate healing processes built into it. A medical doctor is someone who seeks to create a better environment for the body to heal itself. This may be done chemically with medicines, physically with exercise and/or physiotherapy, surgically by removing tissue, environmentally by making an appropriate change in our surroundings, and emotionally by inspiring confidence and health. Despite these interventions, the human body is subject to many factors beyond our control, such as 'incurable' diseases, accidents, and violence from other human beings and from the environment

When Jesus began his ministry following his baptism, he proclaimed the coming of the kingdom of God. In demonstration that the kingdom was near, he healed the sick. The parallel to this event in the Old Testament was the escape of the Children of Israel from the Egyptians through their 'baptism' in the waters of the Red Sea. In their new-found freedom, they were immediately challenged by water that was too bitter to drink; see Exodus 15:22-27. The people grumbled against Moses, who was told by the LORD to throw a particular piece of wood into the water to make it sweet and drinkable. The LORD continued to make it clear that he sought a relationship with his people so that they would keep his commands and decrees. In return, he promised to keep them free from the diseases he brought on the Egyptians (Exodus 15:26). The LORD heals his people, that is, people who accept his rule. Indeed, many of the laws and ordinances he subsequently gave to the Israelites were designed to maintain the physical, mental and spiritual health of the community. God promised that when the Israelites reached the Promised Land he would sustain them by healing sickness, blessing the wombs of the women, and enabling people to live to a ripe old age. Sickness and disease were not to be part of God's kingdom: those who were sick were healed and guaranteed all the resources they needed to live the life God gave: otherwise, they had no place in the community. For example, people with contagious skin diseases such as leprosy had to live outside the camp. However, although healing and wholeness were necessary qualifications for entry into the community of God's people, it was God who healed and made whole.

Jesus came to reveal the Father. When Jesus healed, he called out faith in the person seeking healing and in those who heard him and saw what he was doing. Sometimes he used material aids in healing, such as water. For example, he mixed his saliva with dust to make mud which he then put on a blind man's eyes; the man was told to wash it off in the Pool of Siloam. Even though the man had been born blind, he was able to see for the first time after washing (John 9:1-7). On another occasion, Jesus was at the pool of Bethsaida in Jerusalem (John 5:1-9). A great number of invalids and others who were seeking healing lay around the pool, waiting for the surface to be disturbed 'by an angel'. The first one into the pool after the disturbance was healed of sickness. While at the pool, Jesus came across a man who had been disabled for thirty eight years. Jesus asked the man whether he wanted to get better. The man complained that every time the surface of the water was disturbed, somebody else managed to get down into the water before him. Jesus took this reply as a sign of his desire to be healed, and healed him.

Much human illness and sickness is caused by bacteria and viruses that exist in the environment or are passed from one person to another through human contact. Personal hygiene has been increasingly recognised as a key factor in limiting the spread of disease. In particular, washing with water provides an important means of removing dirt and grime from the skin of the body, and particularly the hands. We learned as children to wash our hands in fresh water from the tap after going to the toilet, a luxury that many people in the developing world do not have.

In Chapter 11 above we explore the use of water in ritual washing related to diseases or biological pollution, such as mildew. Here, we focus on the significance of water in connection with leprosy or Hansen's disease[130]. This is a contagious skin disease which wreaks havoc with people's lives. The disease attacks and kills cells on the surface of the body, eating into the flesh, leaving a white residue and killing the nerves, so that the sufferer cannot feel anything. It spreads over the body and destroys fingers and toes, not so much because of the disease itself, but because the sufferer is not able to feel when he damages them. There was no cure for leprosy until the twentieth century. Before this time, communities had strict rules to minimize the spread of the disease: those affected were banished from the community, restricted to living outside the village or town, and ordered not to come within 50 meters of any other person who did not have the disease. Families were torn apart, and those with leprosy were never able to live with their loved ones again.

In the Bible, the healing of lepers and their acceptance back into their communities was usually associated with water. A famous story concerns Naaman, commander of the Aramean forces in the days of the prophet Elisha. Naaman was a very successful soldier, but he had leprosy. He learned about the prophet Elisha and his miraculous powers from a young Israeli girl who had been captured by his army, and he asked his king to arrange for him to meet Elisha. So the king of Aram wrote a letter to the king of Israel, asking him to heal Naaman, without referring to Elisha. The king of Israel, having no such healing power, thought that the Aramean king wanted to reopen hostilities between them. Fortunately, Elisha heard about the letter and asked that Naaman should come to his house. Naaman went, expecting the prophet to perform some dramatic gesture over him; instead he received a message from Elisha's servant to go and bathe seven times in the river Jordan. Naaman felt insulted. For him, the rivers of Damascus were far better to bathe in than the piddling little Jordan River, and he went away angry. However, he had some faithful servants who persuaded him to do what Elisha said. He bathed himself seven times in the Jordan and was healed. Just as seven is the number of perfection, Naaman was made perfectly clean. As a leper, he was cleansed through washing in water. The Mosaic Law also required washing in water to cleanse the cause of the disease; see Leviticus 14.

Jesus encountered a number of lepers during his ministry. On one occasion, there were ten lepers who met him as he was going into a village. They shouted to Jesus to have mercy on them. When he saw them he told them to go and show themselves to the priest. The implication was that they would be cleansed from their leprosy. They must have obeyed Jesus' command to go in faith that they would be healed. Immediately as they went, the leprosy had disappeared. All of the men must have shouted out for joy. They did what was uppermost in their minds. Nine of them ran off to the priest to get his permission to return to their families and friends; only one came back to praise God for what he had done for him, and to thank Jesus. This one man ran right up to Jesus and fell at his feet. Jesus pointed out the irony that the only former leper to come back and give thanks was a Samaritan, not a Jew. On another occasion, a leper approached Jesus, being bold enough to come right up to him (something that he was not permitted to do), and appealed for his help. 'If you will, you can make me clean'. Jesus' anger was aroused against the debilitating disease: this was not the way God had intended the world to be. Jesus repeated the phrase 'If I will?' turning it into a question. Of course he would, and the man was cleansed. There is ultimately no room for disease and sickness in the kingdom of God.

Water can be the means of life and death. Clean water is important in maintaining our health, but (polluted) water in the natural environment can carry with it dangerous diseases. These diseases are caused by micro-organisms that attack the intestines, such as protozoa and bacteria, with viruses, and parasites that include 'worms'. The World Health Organisation points out that 1.8 million children die each year because of waterborne diarrheal disease[131]. This highlights the urgent need for us to take preventive measures to provide all with fresh water for drinking and safe sanitation.

In the field of health, underground springs of water have long been acknowledged to have curative powers, especially by the Orthodox Church. Often the water from these springs originates deep underground and is warm, having been in contact with hot rocks. Additionally, such water can contain traces of various mineral salts. Europe has a number of 'spas' built up around these springs. Some of the most famous spas are Spa in Belgium[132] (the source also of well-known bottled water), Bath[133] in the UK (with its thermal Roman baths) and Bad Gradierhaus in Austria[134]. The latter has a 110 meter long and 30 meter high construction in the grounds, in which spring water trickles down through ash twigs and enriches the fresh air with salt water particles. Patients are encouraged to walk round the structure where the cooling and humidifying of the air makes it easier to breathe. There is little doubt that people are helped by these spas. But whether more benefit comes from being bodily immersed in the spring waters or walking slowly through a generated humid atmosphere or from the considerable sense of well-being induced by the environments of the spas is still not clear.

What these spas and springs do for us however, is to raise the question as to whether water has a spiritual as well as a physical value. As this book highlights, water has a prominent and important part to play in the Bible. Although the Bible refers to the more obvious physical properties and benefits of water, it also introduces a number of psychological, if not spiritual attributes. Ian Bradley in his book 'Water: A spiritual history', explores the ways in which water has been central to the practices of most of the world's major religions, especially in ceremonial washing, healing, protection from evil and initiation rites. Wells in particular received much religious attention. Pagan healing wells were adopted by the early Christians in England and Wales, with each well dedicated to a local saint and many becoming places of pilgrimage. Water-based healing centres were identified with spas; that is, significant source of fresh water, which took their identity from the Belgian town of Spa. Although Spa water was discovered in 1326[135], it was not until 1777 that the word 'spa' was used for a 'place or resort with a mineral spring'. Meanwhile, many wells were closed during the Reformation. During the seventeenth century, various people developed and promoted water cures, including drinking and bathing in water from particular spas. Spas in England at Bath, Buxton, Harrogate, Cheltenham and Malvern Wells, among others, became the focus for certain water cures that encouraged 'water drinking, socialising, and promenading in peaceful surroundings'. Some of the doctors associated with these cures became famous society physicians, and the corresponding towns and cities have retained their spa image, even though the medical focus has greatly diminished in the light of modern, science-based medical treatments.

Within Christianity, the spiritual aspect of water has meant that it is now regularly sprinkled on the people for cleansing in some churches, and plays a prominent role in baptism throughout the world-wide church. Baptisms in the early church were often held in flowing water, such as a river or the sea, so it was difficult to appreciate the blessing of a specific quantity of water. In the Middle Ages, water blessed for baptism was often kept in a font with a cover that was locked. This was to prevent the water being stolen and used for unauthorized magic practices. Water is still used by Catholics, Anglicans, Eastern Orthodox churches, Lutherans, Baptists and others for baptism, the blessing of people, places and objects, and as a sacramental means of repelling evil. A priest can bless water in a vessel such as a font or stoup, and thereby make it available to others. Today, Catholics are encouraged to dip their hand in holy water in a 'stoup' at the entrance of the church, and to make the sign of the cross. On Sundays, the liturgy of the mass may begin with the ninth century rite of blessing and sprinkling with water, called aspersion, which comes from the Latin for 'to sprinkle': asperges. The water is sprinkled on the congregation using an aspergillum, which is a form of brush (or branch). Whereas the use of water that has been blessed is more likely to be in a Catholic church, it is gaining recognition in higher forms of Anglicanism. Undoubtedly water has a definite and more profound spiritual role than many of us may be aware of. Ian Bradley's book is timely in that it explores the various ways in which water has had a spiritual impact down the centuries, especially in Europe and the UK[136].

Resource questions:

What stories have you to tell about therapeutic experiences of water?

How can we make best use of water for healing?

32

Water and the Cross

> 'The Church's one foundation
> Is Jesus Christ, her Lord;
> She is his new creation
> By water and the word'
>
> Samuel John Stone, in 'Lyra Fidelium;
> Twelve Hymns on the Twelve Articles of the Apostles' Creed.'
> See http://en.wikipedia.org/wiki/
> The_Church's_One_Foundation

Read John 19:16b-37

From the beginning of the Old Testament scriptures, the prophets sought to penetrate the secrets that God had hidden from the foundation of the world, namely of his final plan to redeem human beings. The Cross was the culmination of those centuries of prophecies. So it was somewhat poignant that Jesus should be transfigured with Moses and Elijah, the two great prophets of the Old Testament, with their main topic of conversation being what Jesus would accomplish at Jerusalem. The cross was uppermost in Jesus' mind. Through the transfiguration Jesus allowed his three most trusted disciples an insight into the reality of his being. Their experience helped them preserve their faith, even though they deserted him, and the others (not John) were unable to stand at the cross as Jesus suffered. His own disciple, Judas, betrayed him; and the others left Jesus to the illegal and subversive actions of the religious leaders who condemned Jesus before Pilate, and to the Roman governor's complicity in granting their request that Jesus should be crucified, when it was obvious to him that the prisoner had done no wrong.

After being tortured by the Roman soldiers Jesus was led out to Calvary to be crucified. Business for the soldiers was brisk that day, with two other prisoners to be put to death as well. They made

Blood and water flowing from Jesus' side

the prisoners carry their crosses to a hill outside Jerusalem on one of the main routes into the city. They forced them to lie down on the crosses while they stretched out their arms on the cross-pieces and hammered nails through their wrists to fix them to the wood. Then they pulled teach prisoner's feet together and hammered a long nail to fix them towards the bottom of the upright. Once this was done the soldiers lost no time in hauling up the crosses with the three prisoners hanging down from the cross pieces, and slotting the uprights into prepared holes in the ground. It was all in a day's work for the soldiers. All three prisoners were crucified on the hill, with Jesus in the middle.

The pain in wrists and feet from the nails was intense, but that paled beside the increased pressure on their lungs as they hung on their crosses. The so-called religious leaders took turns at shouting taunts at Jesus to come down from the cross. Meanwhile Jesus complained of thirst. They soaked some vinegar on a sponge and raised it on a pole to his lips. Perhaps they were expecting some dramatic action from him. Finally Jesus cried out 'it is finished' and breathed his last. It had scarcely taken six hours for him to die. The others struggled to raise their bodies on wounded feet so that they could breathe growing weaker. The chief priests wanted the bodies off the crosses as the Sabbath approached, so they petitioned Pilate to end the crucifixions. Pilate asked for a report from the centurion in charge and found out that Jesus was dead, but the other two were still alive. Pilate was surprised that Jesus had died so soon. He ordered that the legs of the other two should be broken so that they would die more quickly. This was done, and a soldier pierced the side of Jesus just to make sure he was dead. He reported that both blood and water flowed from the wound.

Tears of remorse from Judas and Peter, tears of grief from the women at the cross, vinegar to drink, blood and water streaming from Jesus' side, his body washed quickly with water and deposited in a stone tomb. Water was at the heart of the event that was the Cross. No wonder that John, in writing his Gospel and epistles, consistently states that we must be born again of water and the Spirit. Jesus was the rock which was struck by the lance of a Roman soldier (compare the rock struck by Moses with his rod; see Exodus 17:1-7 and Numbers 20:1-13), and out came the water and blood which would give life to the people of God. Blood and water are seen as being symbolic of the two sacraments of the Church: Eucharist and baptism. The water was the physical sign of the coming and soon to be indwelling Holy Spirit (John 4:3). Zechariah refers to the spring of water flowing from a crucified Messiah; see Zechariah 12:10-13:1. The blood is symbolic for those who come for forgiveness (in faith) to the altar of Calvary. We can draw near to God 'with a sincere heart in full assurance of faith, having our hearts sprinkled (with the blood of Christ) to cleanse us from a guilty conscience and having our bodies washed with pure water'; see Hebrews 10:22.

Some physicians suggest that the separation of blood and water are a symptom of a ruptured heart. This would provide some indication of how Jesus gave up His spirit (John 19:30) and died. In this instance he, as creator, burst His own heart and died. Jesus could be said to have literally died of a broken heart out of his love for those he came to save. The important conclusion from this observation is that Jesus was positively identified as dead before His burial. Some theologians go on to suggest that the blood symbolizes the salvation of the individual, whereas water represents his or her sanctification.

Then on the third day (counting the day that he died as the first day) Jesus was raised from the dead. By all accounts, his resurrected body was made up of the same chemical elements, including water, as before. But his body was not restricted to the laws of this physical universe; see John 20:19. Precisely how that happened we do not know. Suffice it to say, that in Christ our physical humanity is in some sense part of the Godhead, and water is a crucial part of that! Jesus took our flesh to God's throne. There has been, it appears, a profound re-integration of the physical and the spiritual worlds.

Reference:

Bradley I, (2012) '*Water: A spiritual history*'. Bloomsbury

Resource questions:

What does the death of Christ mean for you?

Are you aware of the Holy Spirit within you as a spring of water providing eternal life?

33

Waters of baptism

'If you wish to drown do not torture yourself with shallow water'

Bulgarian proverb

'The last recorded use of a ducking stool anywhere in England was in Leominster (Herefordshire) in 1809 when Jenny Pipes was paraded through the Town on the stool and ducked in the Kenwater by order of the Magistrates. On a later occasion, in 1817, Sarah Leeke was paraded, but could not be ducked as the water level was too low.'

From the website of Leominster Parish Church

www.leominsterpriory.org.uk

Read John 3:1-8

The verb 'to baptise' is derived from the Greek word which originally meant 'to drench' or 'to drown', and was also the word for a sinking ship. There are therefore the notions of being swallowed up, of being cut off from life, of dying, or being transformed through being immersed in some form of liquid. The act of immersion is physical, during which the initiate experiences the actual physical death by being immersed in the water. Alternatively, the physical act may be symbolic of a much more profound and spiritual or psychological transition, in which consequently, what baptism conveys is the sense of a radical transition from one state to another; the transition for the initiate can be as dramatic as physical or spiritual death.

The early Church Fathers identified five forms of baptism: (i) the baptism of Moses when the Children of Israel escaped 'through' water from the Egyptian army and is therefore purely allegorical; (ii) baptism in water with repentance, as with John the Baptist; (iii) Christian baptism in water as 'an outward and visible sign of an inward and spiritual grace' through the Holy Spirit in the name of the Trinity;

(iv) baptism in blood through martyrdom; and (v) baptism in tears of repentance (the gift of tears) for one's sins committed since (Christian) baptism[137]. Water was considered so important as the medium for baptism that Ignatius of Antioch said in his Letter to the Ephesians that '(Jesus) was born, and he submitted to baptism so that by his passion he might sanctify water'. [138]

Indeed, the most prominent place that water has in the Christian tradition is in baptism. Jesus disturbed Nicodemus by stating that no-one can enter the kingdom of God unless he or she is born of water and the Spirit (John 3:5), implying that Christian initiation involves both water and the Holy Spirit. Matthew confirmed that Christian baptism is carried out in the three fold name of the Father, Son and Holy Spirit. Therefore, baptism is a major sacrament of the church, through which it initiates its members.

But why and how are water and the Holy Spirit involved? The roots of Christian baptism can be seen in the use of water, as the primordial substance for creation, for the symbolic (or even practical) washing away of impurities in preparation for the individual's approach to God in the sacrificial system, and in the initiating rites of Jewish proselytism. Christians see baptism prefigured in the survival of Noah and his family through the flood (1 Peter 3:21) and the crossing of the Red Sea by the Israelites in their escape from Egypt (1 Corinthians 10:2). Just as everything was created by the word of God out of water and by water (2 Peter 3:5), so water used in baptism is the symbol of the re-creation of the individual by the working of the Holy Spirit in his or her heart and life.

We first read about baptism in the Bible when John the Baptist was preaching, calling his fellow Israelites to repentance. He was an uncompromising prophet, who was born to his parents, Zechariah and Elizabeth about six months before the birth of Jesus. Elizabeth was related to Mary, Jesus' mother. Living in the desert of Judea some 30 years later, John challenged people to a radical repentance of heart and mind from sin, a repentance which was sealed in baptism by immersion in water. By his preaching, John attracted many ordinary men and women from the towns and villages round about. The people were expecting the imminent appearance of the Messiah, who had been promised for so long in the Jewish scriptures, and they were convicted by John's message. So they flocked to John, wanting to be ready for the Messiah's coming, and John baptized them in the River Jordan.

John had many followers; so many that some questioned whether John was himself the Messiah. When asked, however, John emphatically stated that he was not, saying that he was unworthy even to untie the Messiah's sandals, something only the lowest servant in the household would do. John also raised expectations by declaring that the Messiah would baptize not with water as he did, but with fire and the Holy Spirit; his ministry would be of a different order to Johns.

Then Jesus came to John in the desert to be baptized. He sought baptism, not as a sign of repentance for sin but as an initiation of ministry. When John saw Jesus approaching, he knew that the Messiah was coming to him,—but for baptism? How could this be? If anything, he (John) should be baptized by Jesus. But Jesus was adamant; and as he came up out of the water following his baptism, John recorded that he saw the Spirit settle on him as a dove; see Luke 3:21. A voice from heaven said 'This is my beloved Son with whom I am well pleased'. John was left in little doubt that Jesus was 'the Lamb of God who takes away the sin of the world', and he told his disciples so; see John 1:32. As a result, two of John's disciples, one of whom was Andrew and the other possibly the Apostle John, left John the Baptist and followed Jesus; see John 1:35-40

Later in his ministry, Jesus made baptism with water in the name of the Trinity the necessary initiation rite into the church; see Matthew 28:19. From the discussion with Nicodemus (John 3:5-8) we can see that Christian baptism is an outward, visible and sufficient sign of the indwelling Holy Spirit. But it is not a necessary sign, as the encounter of Peter with the Gentiles in Cornelius' house shows; see Acts 10. However, as regards initiation into the church, baptism is necessary (and sufficient)!

'Whoever believes and is baptized will be saved, but whoever does not believe will be condemned' (Mark 16:16). The theology of baptism is addressed by Paul in his letter to the Romans. Paul explains that in going down into the water in baptism believers identify themselves with Christ in his burial. Similarly, when they come up out of the water, they share in the resurrection of Jesus; that is, in his rising to new life. Elsewhere, Paul speaks of his own experience of being crucified with Christ, and of being buried with him. Liturgically, Christian baptism involves two movements: a 'turning away from' and a 'turning towards'. Candidates for baptism declare that they repent of their past

The baptism of Jesus

lives, they reject the works of the devil, and they resist evil. In contrast they acknowledge Christ as their LORD and Saviour, and that he is the Way, the Truth and the Life. The Catholic, Anglican and Episcopalian churches generally practice baptism of infants within a Christian family, with parents and godparents making the promises for the child in the expectation that, encouraged by them, the child will one day take on the baptismal promises for him or herself. Usually these churches expect the child to do this when confirmed by the bishop.

Some Orthodox churches and religious sects have an even stronger association with water in baptism. Lady Drower was an anthropologist who recorded the rites of a number of churches and sects in the Near to Middle East. She commented that water was the medium for a pre-Christian, secretive sect in Iraq and Iran called the Mandaeans 'which most fully expresses the mystery of being, or of the Being which is semi-personified as the Great Life'. Similarly; a river is the equivalent of the 'heavenly Jordan flowing in the world of light'[39]. Immersion in water, therefore, is an act which purifies, revitalizes and protects. The spiritually dead—and pollution is a form of spiritual death—can be reborn through baptism. Baptism becomes a *rite de passage*: it corresponds not only to the babe's transition from the darkness of the womb to the outer world, but also to the passage of the soul from the world of matter into that of spirit'. Each candidate for Mandaean baptism[140] experiences the rite individually, but subsequently the candidates take part in a ritual meal (the equivalent of communion) together. In particular, they ate sacramental bread which was baked on a brazier, and they drank water from the flowing river in which they were baptized.

An early Church training document, (the Didache), requires that Christian baptism, where possible, should be carried out in flowing water in order to align the water of baptism more closely with the 'living water' that is the gift of the Holy Spirit.

Because the Greek word 'baptism' carries meanings such as 'drowning' or 'drenching', it would appear that baptism should be administered by immersing the candidate in water. But with the acceptance of whole families into the early church, and therefore small children being baptized, the church came to pour water on the head of the candidates, especially where water was in short supply. As Lady Drower has pointed out, baptism is practiced in very different ways in the diverse parts of the church world-wide as well as other religious sects[141]. But whatever the theological interpretation or cultural background, water, subject to the word of God, is at the heart of the practice of Christian baptism.

References

Didache, early Christian teaching document

Drower, E S, (1956) *Water into Wine: A study of ritual idiom in the Middle East.* John Murray

Resource questions:

Jesus baptizes with the Holy Spirit and with fire. Have you been baptized in water and in the Holy Spirit? What difference do these baptisms make to your life? What are the benefits and disadvantages of being christened (baptized) as a baby, or as an adult?

34

The gift of tears

'When it is genuinely spiritual, 'speaking with tongues'
seems to represent an act of 'letting go'—the crucial
moment in the breaking down of our self-trust, and
its willingness to allow God to act within us. In the
Orthodox tradition this act of 'letting-go' more often
takes the form of the gift of tears.'
'The Orthodox Way' by Bishop Kallistos Ware,
A. R. Mowbray & Co. Ltd. 1979

'Oh, that my head were a spring of water and my eyes a
fountain of tears!
I would weep day and night for the slain of my people.'
Jeremiah 9:1

Read John 11:30-44

Tears are usually a sign of deep emotion, though the precise nature of
the emotion may be difficult to interpret: we can weep for joy, grief,
sadness, pain or relief, or simply because we are slicing an onion! Jesus
calls us all to weep with those who weep; that is, to be identified with
others in their distress, to side with them over the reason for their
tears and to encourage them as they work through their emotions and
memories. He also teaches in the second beatitude that 'Blessed are
those who mourn (or weep), for they shall be comforted' (Matthew
5.3). This particular verse was formative in directing the Early Church
Fathers to identify tears as a gift from God when they are focused not
on mourning for friends or relatives, or sadness over a mishap, or even
the sufferings of Christ, but on 'deep regret felt in the depths of the
heart on having sinned'; see Hausherr (1891)[142].

The fluid released by the tear ducts in the eyes is critical for
keeping the cells on the surface of the eyes wet. This is because these
cells are alive, unlike the cells on surface of the skin that covers our

bodies. Blinking moves fluid around the eye and ensures that they do not dry out. When we cry, other glands come into play. These generate a different fluid onto the surface of the eye which flushes minerals and hormones associated with depression and stress from the body. So much fluid can be generated that the ducts which drain the fluid away from the eyes cannot cope, and excess fluid collects along the lower eyelids and flows over the edge or down into the corner of the eyes. As mentioned above, a variety of both physical and emotional reasons cause tears to be released. One reason could be an irritant in the air such as tear gas! Another could be grief over the death of someone close to you, or exhilarating joy over a new baby being born. However, there can be occasions for many when something takes them completely by surprise, and causes tears to come to their eyes each time they experience the associated emotion.

For example, the joy and exhilaration of standing head-on into the wind on the top of a mountain or high hill overlooking an impressive view may lead to tears. Alternatively, tears may reflect an observer's sadness for the intransigence of a dysfunctional group of people who refuse to listen to each other and instead shout each other down.

In the Bible, tears are referred to about 200 times. Hagar, the mother of Ishmael, wept in despair after Abraham's wife, Sarah, treated her so harshly that she ran away; Genesis 16:6. Esau wept before his father, Isaac, because his brother Jacob had stolen his blessing. Joseph wept secretly when he observed his brothers, who were unaware of his identity as they discussed their guilt at having sold him years before into slavery. Later, he wept openly before them as he revealed his identity. The Children of Israel, as slaves in Egypt, wept bitterly before the LORD, who heard their cries and called Moses to their rescue. During the Passover Seder, the Jews continue today to remember the tears of bondage in Egypt as they sip the salt water in the cup of tears. Hannah wept at Shiloh before Eli as she prayed for God to give her a son. David wept at his separation from Jonathan when Saul was seeking to kill him. He also wept before God for his critically ill son, born to Bathsheba; and later in (old) age he wept following the death of his son, Absalom, who had tried unsuccessfully to wrest the throne from his father. Hezekiah, when dying, prayed with tears for recovery; God subsequently extended his life by fifteen years. Several psalms refer to tears being shed in prayer when there is failure to keep the law (see Psalm 116:8 'You O LORD, have delivered my eyes from tears'. Psalm 126:5, 6: 'Those who sow in tears will reap with songs of joy'. Psalm 130:1: 'Out of the depths I cry to you O LORD'). The prophets identified with people they were sent to through their tears. Many tears were shed by exiles from Israel and Judah, both in the lands they were taken to, and when they later returned to Jerusalem. Mourners put their tears in bottles, even wearing them round their necks (Psalm 56:8).

People often wept in Jesus' presence, and Jesus on occasion wept before others. Once, a woman came up behind him when he was reclining at table in the Pharisee Simon's house; see chapter 15 of this book. The woman was rying. She bathed his feet with her tears and dried them with her hair. Jesus knew that her tears expressed her great love and her sorrow for her sins. Later in the gospel Jesus heard that his friend Lazarus was ill, but he deliberately delayed going to Bethany until he knew that Lazarus had died. When Jesus approached the tomb where Lazarus had been laid, he found Lazarus' sister, Mary, and other Jews weeping. Jesus was greatly disturbed in spirit and deeply moved. Then when he was invited by the Jews to come and see the tomb, he wept; see John 11:35. The Jews explained his tears as being due to his great love, but he could also have been weeping because he was observing the despair of the people in the face of death, or because he was angry that death had invaded the lives of his friends. Spurgeon claims that Jesus wept because 'this was his method of prayer on this occasion'. When he was approaching Jerusalem just before his arrest, trial and crucifixion, he saw the city from afar and wept over it. He lamented the city's ignorance of what would bring (it) peace; see Luke 19:41-42. Jesus prayed regularly to his heavenly Father. In the Garden of Gethsemane, he prayed 'with loud cries and tears . . . (to) the one who was able to save him from death' (Hebrews 5:7-8a).

In the gospels we read of mourners weeping for the widow of Nain's son (Luke 7:11-15), for Jairus' daughter (Luke 8:40-56) and for Lazarus (John 1:41-44). Peter wept in shame when he realized that he had denied his master (Luke 22:62). On Easter Sunday morning, Mary Magdalene wept at the tomb where Joseph of Arimathea and Nicodemus had put Jesus' body, thinking that the body had been deliberately removed; see John 20.10-18. Later we read that the Apostle Paul was no stranger to tears. In saying farewell to

They that sow in tears shall reap in joy

the Ephesian church leaders, Paul reminded them that he had never stopped warning each of them night and day with tears (Acts 20:31). He also recalled the tears of others, including Timothy (2 Timothy 1:4)

References to tears continue into Revelation. John wept when no one was found worthy to open the seal; that is, until the Lion of Judah appears; see Revelation 5:4. Later the merchants wept as they beheld the demise of Babylon the great city; see Revelation 18:11. Finally, God assured his people there will be no more tears in heaven, indeed God will himself wipe all tears from their eyes; see Revelation 21:4.

I was at university in the early 1960s, during the era of the Death-of-God movement, among other cultural changes[143]. This turned out to be a prelude to the charismatic renewal of the Christian church world-wide, which has been going on ever since. John Richards (2001) identifies a number of profound personal consequences of this renewal, including a deeper relationship with Jesus, a greater awareness of God as Father, the overwhelming desire to praise and worship the triune God, a stronger abhorrence of personal and corporate sin, a revitalized love of Scripture, a heart-felt ambition for Christian unity and deep sorrow at our disunity, a conviction and determination to share the Gospel with others, and an eagerness to work out the social implications of the Gospel. He states that it is in this context tears are shed which are 'associated, not with human passions, but with the experience of God. Even their physiological aspect (of those shedding the tears) manifests this fact. They (the tears) flow without strain or effort, without violent sobbing or the contortions of the face muscles'.

Many Christian writers from different traditions and churches acknowledge such an experience as the 'gift of tears'. A common interpretation of the experience is that it makes individuals profoundly aware of their distinctiveness from a holy God, being overwhelmed by the gravity of human sinfulness while longing for a deep unity with him. It is as though individuals are identified with the Jews exiled in Babylon, weeping for their restoration to Jerusalem while overwhelmingly aware of the sins that brought about their exile; see Psalm 137. This is consistent with the statements proposed by Tim Keller as a summary of what God has done for us in Christ: 'In and of myself I am more sinful than I ever dared to believe, but in Christ I am more loved than I ever dared to hope.'[144] As they have done for countless Christians down the ages, our sins and sinfulness cause us to lament our state; by rights we should be alienated from Christ. Yet despite our sins, with Christ and in Christ, God accepts us and loves us more than we can hope or understand.

However, we should not assume that tears necessarily provide a ready access to God. 'Crocodile tears' hide insincerity, whether realised or not. The LORD saw through the tears of the people of Judah who had broken their covenant with God; see Malachi 2:13-16. The prophets declared that as well as tears there needed to be changed attitudes towards the widows and orphans, the poor and the infirm.

Early Church saints talked about 'compunction', which is the name for those 'emotions that come from supernatural thought'[145]. There is a duality about a person's relationship to God. Out of his infinite love for man, 'God, when he erases sin, does not leave a scar, does not even allow a trace to remain, but with health he restores even beauty'. Our transgressions and sins lead us to mourn unceasingly; while in our hearts we express our thankfulness to God for all that he has done for us in and through his one and only Son Jesus Christ. Maggie Ross, in her book 'The Fountain & the Furnace', has made an extensive journey into the psychological and spiritual aspects of the gift of tears. She sees poverty and solitude as reasons for tears; but she also looks at the psychology of tears and the ways in which different writers have interpreted and advocated the gift of tears. She emphasises that tears are essentially for our sins that continue to bedevil us; yet these tears of sorrow turn to expressions of joy in the God who suffers with us.

Eastern Orthodox Christians have a number of ways of describing this gift: the way of tears, the prayer of tears, the gift of tears, holy sadness, tears which illuminate, and weeping without ceasing. They regard the gift as such a central part of the spiritual life of the Christian that they call it 'the second baptism'; that is, the waters of (the first) baptism deal with past sin and initiation into eternal life, while the waters of the gift of tears relate to God's washing away of our present sins. The gift of tears enables a right emotional response to God, a change in priorities, a freeing up of crying through the healing of emotions, an encounter with our real selves, and a deeper awareness of evil and corruption in the world. It is found in all parts of the church where believers take God seriously and seek to love and serve him.

References:

Richards, J (2001) *'Tears—Gift of the Spirit?'* Available at www. helpforchristians.co.uk

Ware, Bishop Kallistos (1982) *The Orthodox Way*, Bishop Kallistos Ware

Ross, M (1987) *The Fountain and the Furnace: The way of tears and fire.* Paulist Press

Hausherr, I (1891) *Penthos: The doctrine of compunction in the Christian East.* Cistercian Publications

Resource questions:

Can you identify the gift of tears in your experience?

What benefits does the gift of tears provide to us as individuals and to the church?

35

Down to the sea

'He stretches out the north over empty space,
And hangs the earth on nothing.
He binds up the waters in his thick clouds,
And the cloud is not burst under them.
He encloses the face of his throne,
And spreads his cloud on it.
He has described a boundary on the surface of the waters,
And to the confines of light and darkness.
The pillars of heaven tremble
And are astonished at his rebuke.
He stirs up the sea with his power.'

Job 26:7-12

'The river is within us, the sea is all about us;
The sea is the land's edge also, the granite
Into which it reaches, the beaches where it tosses
Its hints of earlier and other creation'

'The Dry Salvages' in 'Four Quartets' by T. S. Eliot,
published in 1944

Read Acts 27

The Hague has a number of fine museums, displaying historical objects from the Netherlands and all over the world. Some of the best exhibitions are of paintings and artifacts made in The Hague and its surroundings. One particular painter who had considerable effect on the city and its life is The Hague School painter H. W. Mesdag. His delight was to paint seascapes and the characteristic flat-bottomed fishing boats that used to be pulled up out of the water onto the beach at Scheveningen by teams of horses[146].

When entertaining guests to our home in The Hague, my wife and I often take our visitors to the Panorama, which Mesdag, his wife and some friends from The Hague School painted within a few weeks in 1888. We enjoy our guests' surprise when they suddenly emerge from darkened stairs on to what seems to be the top of a dune in Scheveningen, surrounded by a circular panorama showing the beach, with fishing boats, fishermen and their wives, cavalry and holidaymakers on the one side, and the dunes and village houses, churches, pumping stations and canals on the other. The effect is so direct that the viewer can readily imagine that the sea is actually breaking on the beach with the piercing cries of the seagulls overhead. The scene is peaceful and hides the harshness of the fishermen and women's lives. Not long after the panorama was painted, a violent storm destroyed many of the boats and people's livelihoods. Money was subsequently raised to construct a harbour in Scheveningen where their boats could be protected and people's lives could be made more secure[147]. The boats no longer had to be hauled up on the beach, and they were replaced by fishing boats with a deeper draught and better safety.

Undoubtedly the sea was a constant threat to people, but it also provided limitless opportunities. Ever since the emergence of the first civilizations on major rivers, there have been those who saw the sea as a means of linking one community with another, and of transporting large quantities of goods more easily than over land. Whereas mariners at first kept close to the coast, they soon began to explore uncharted waters, whether through storms driving them in directions they did not wish to take, or in their desire for adventure. Sailors looked for new lands, new trading goods, new fishing grounds, and new experiences, and they also experienced the capriciousness of the oceans. They endured violent storms, they were driven off course; some never returned home; others met whales or discovered a motley collection of weird fish.

Though Ferdinand Magellan was the first man to sail around the world in a sizeable ship manned by experienced sailors in 1585-1589, it was not until 1967 that Sir Francis Chichester became the first man to sail solo around the world[148]. This feat, like the climbing of Mount Everest in 1953 by Hillary and Tensing[149], has been repeated many times since then, with women, disabled people and even teenagers achieving it. It seems that once something has been shown to be achievable, others rapidly follow.

The Jews retained in their psyche a deep suspicion about the Great Sea (that is, what we know today as the Mediterranean). It was other nations who became sailors, deep sea fishermen and traders, and who developed the economy of the region, even in Solomon's day. The Psalmist applauded those 'who go down in deep water' because they put themselves (as far as he was concerned) in great danger (Psalm 107:23-32). They floated on the surface of the ocean in wooden ships, powered by the wind or the oars of slaves, steered by a rudder and were therefore at the mercy of strong winds and mounting waves, tidal currents and submerged rocks. Occasionally they experienced extreme storms when waves threatened to overwhelm them, currents were so strong that they were helpless to go against them or to escape their grasp, and where rocks impeded their progress and were a danger to the integrity of the structure of the ship. Striking a rock will almost always result in a hole in the hull of a boat or ship, leading to the inflow of water which reduces buoyancy. A large enough gash in the hull can mean that the ship will sink quickly. Perhaps the most famous example of this is the Titanic, which was thought by its designers and owners to be unsinkable; after striking an ice-berg, sealed units internal to the ship were overtopped by seawater, and the ship sank in a matter of hours with huge loss of life[150].

Sailors in the Mediterranean were not threatened by icebergs, but nevertheless waves could be huge, and depending on the size of the ship relative to the waves, the structure of the ship could be put under severe stress. This was enough to make even experienced sailors despair of life itself. Hardened sailors cried out to God in their trouble. This was the experience of the Apostle Paul, when he was taken as a prisoner to Rome. At the end of the Acts of the Apostles, Luke wrote about the shipwreck that he and Paul experienced as the climax of his book. This story parallels the arrest, trial, suffering and death of Jesus as the climax in Luke's first book, his gospel. His two volumes bear all the marks of a serious historian determined to present the information of historical happenings according to the prevailing culture and traditions as honestly and transparently as possible.

We read about Paul's journey to Rome in Acts 27 and 28. Paul, with his travelling companion, Luke the physician, was taken on board a

Paul's shipwreck

ship that sailed from Adramyttium in Caesarea to ports on the coast of the province of Asia. At Myra in Lycia, the centurion in charge of Paul found an Alexandrian ship sailing for Italy. We know it was a sizeable sea-going vessel, because it had 276 people on board, including other prisoners, soldiers and the sailors. They made slow headway against strong winds, and came to a harbor, called Fair Havens, near Lasea on Crete. Winter was approaching and conditions could now be treacherous, so the owner of the ship wanted to make for Phoenix, another port on Crete, to spend the winter. Paul warned the centurion against sailing, stating that the outcome would be disastrous for the ship and its cargo, let alone human lives. He was over-ruled, and with a gentle south wind they set sail for Phoenix. However, a severe storm with hurricane force winds blew up; the sailors could not control the ship and it drifted towards the nearby shore. The sailors did everything they could to make the ship secure. They lowered the sea anchor. They went with the wind and currents. The storm lasted for several days, and with each passing day their hope of being saved diminished. Then Paul told the people on board that an angel of the God he served stood beside him the night before, and told Paul not to be afraid. He was to stand trial before Caesar, and God had graciously given him the lives of all who were sailing with him. Paul had faith in God that this would happen, but they had to run the ship aground on a sandbar. On the fourteenth night of the storm, the sailors sensed that they were approaching land. Disaster was imminent. Concerned for their own lives, the sailors attempted to escape in the lifeboat, but Paul knew that the lives of all those on board would depend on the sailors' skills. So he persuaded the centurion, with whom he had built up a good relationship, to have his men cut the lifeboat adrift. Paul continued to assure everyone that they would be saved, and persuaded them to eat some food. When daybreak came the sailors set sail for the beach which they could see beyond the surf, but the ship ran aground on a sandbar as Paul had foretold, and began to break up. The soldiers planned to kill the prisoners to stop them escaping, but the centurion wanted to save Paul, so he would not let them carry out their plan. He ordered those who could swim to jump overboard and head for the beach: the others were to get there on planks of wood from the ship. They all made it safely to the shore, where they discovered they had landed on Malta, an island south of Italy.

After some amazing events on the island, which endeared Paul to the islanders, the centurion and the soldiers continued to escort Paul and his friends to Rome. The shipwreck obviously made a big impression on Luke. It is not surprising that shortly after writing about the shipwreck, which featured so prominently in his and Paul's experience, he brought his account of the Acts of the Apostles to an end: there was nothing quite as dramatic as the shipwreck again in Paul's life. It is as though water and ocean had threatened to halt the progress of the gospel, but through the grace of God, Paul's witness to the coming of the kingdom of God through the death of Christ, which had been confirmed through his resurrection, was continued right at the heart of the Roman Empire.

If ocean waves can dominate the outlook of a sailor at sea, their magnitude can be even more frightening to those on land, esecially when a tsunami strikes the coast. Tsunamis are caused by earthquakes under the ocean bed. A small rise or fall in the sea bed leads to a correspondingly small rise or fall in the whole of the water above it. Although the vertical change in the water level seems small, the amount of energy transmitted to the water can be enormous, and this is transferred to a horizontal motion of the sea surface as a very long wave travelling across an ocean 1000 m deep, say, with a speed of about 350 km/hr. Although the wave is barely perceptible on the open ocean, the water level greatly increases as it approaches the shore. It is as though the energy in the wave, originally dispersed in the ocean above the earthquake zone, becomes concentrated at the beach. A tsunami wave, which can be as much as 20 m or 30 m high, forms a wall of water that crashes over the beach and the immediate hinterland, sweeping away houses and crops and devastating the coastal zone. Such were the tsunami waves recently recorded along the Indonesian and Indian coasts[151] in December. There were no warning systems in place for the tsunami in the Indian Ocean; the Japanese however, have excellent systems in place to warn of earthquakes and resulting tsunamis, and this, coupled with satellite communications, enabled the world to watch in amazement—and concern—as the devastation along the east coast of Japan took place in March 2011[152].

Europeans should be careful not to think that tsunamis cannot happen to them. In 365AD there was a violent earthquake under the eastern Mediterranean Sea[153] which led to the generation of a particularly catastrophic tsunami that devastated the island of Minoa, and coastal zones in different countries as far apart as Egypt, Tunisia and Greece.

Finally, although the nature of the event was very different, the superstorm Sandy caused enormous damage along the eastern seaboard of the United States of America in October 2012. More than 60 million people were affected by extreme winds and flooding. The winds forced huge waves to travel inland from the coast washing away houses, roads, train tracks and power lines. It took several weeks before power was returned to a number of communities and it took months before the infrastructure was back to normal.

When completing this chapter in November 2013, Haiyan, one of the most devastating typhoons ever observed, with winds in excess of 200 miles/hour, ripped through much of the Philippines, and then Vietnam, leaving thousands dead and hundreds of thousands without homes. It does appear as though those peoples and nations who suffered in the past from natural disasters are increasingly likely to suffer more acutely in the future. The world community has to resolve how best to meet the consequences of these catastrophic disasters in order to minimise suffering and death.

Resource questions:

How well are you surviving the storms of life?

What beliefs and attitudes do you need to survive well?

36

Water in the Middle East

'Fierce national competition over water resources has prompted fears that water issues contain the seeds of violent conflict.'

Kofi Annan, Secretary-General of the
United Nations, 'UN warns of looming
water crisis' BBC News 22 Mar 2002

Read Psalm 65

The Middle East, including North Africa from Morocco to Iran, is arid or semi-arid. The average rainfall across the region is about 750 mm a year. South eastern Arabia has virtually no rainfall, while the region between the Caspian Sea and the Alburz mountains of Iran receive in excess of 1500 mm per year; see Burke (2009). There are three major rivers in the region, each of which are referred to throughout the Bible, namely the Nile, Euphrates and Tigris. Each has its own individual characteristics.

The River Nile competes with the Amazon as the longest river on earth[154]. It is a most unusual river, with its source in Lake Victoria in East Africa, and flowing 4000 km northwards through the Sudan to the Mediterranean to the north of Egypt. Lake Victoria provides water for many countries downstream, but especially Egypt. The main branch of the Nile is called the White Nile, flowing from Lake Victoria. But much of the water in the White Nile is evaporated in an extensive wetland called the Sud. In fact, much of the water that reaches Egypt comes from a major tributary originating in the Ethiopian highlands, called the Blue Nile. Egypt, with its population in excess of 100 million people, could not survive without the waters of the Nile. This dependence has fashioned the politics of the countries in the Nile basin, with Egypt being given absolute rights in international law over the water from the ten countries upstream through or from which the White and Blue Nile rivers flow. Egypt's government recognises its potential vulnerability to interference in the flow of the Nile by other nations upstream, and therefore seeks to secure its water supply by not permitting any artificial modulation of the flow in the river.

The Nile used to flood its delta in Egypt regularly each August, until the construction of the high Aswan Dam in 1970. The Tigris also has some major dams to retain water. Generally it overflows its banks in April, catching out farmers who have sown spring crops. The flows are vigorous, unlike the Euphrates whose sluggish flood waters are dispersed through its lower flood plains.

Burke draws attention to the long history of irrigation and water management practised in the vicinity of these three rivers and elsewhere, going back to 5000 BC. The invention of different means of lifting water such as the Archimedian screw and a water wheel called the noria, together with weirs and dams, canals and qanats, meant that engineers had at their disposal tools to ensure irrigation of arid lands. Besides the ancient civilisations of Egypt, Assyria, Babylon and Persia, the Romans, and then Islam and the Ottomans, left their distinctive marks on the water resources of the region. In the present day, because rainfall is limited, more attention has been given to exploiting the vast underground water resources in the region.

Nobody makes better use of its underground water resources than Israel. Although reasonably well supplied with fresh water, it is unevenly distributed. For example, the average annual rainfall in north-western area of the country is in excess of 900 mm/year, but the extreme south receives only 30 mm/year. Agriculture is feasible only in areas that have more than 300 mm/year. The rainfall feeds the major river, the River Jordan, and three main aquifers that are largely found beneath the West Bank of the river, which is land allocated to Palestine. The upper catchment drains into Lake (or the Sea of) Galilee. The Jordan is fed by three major tributaries, along with springs and the interaction with groundwater. The commercial exploitation of the sources of fresh water and the treatment and distribution of potable water are the responsibility of Mekorot, the Israel Water Authority[155], which also supplies water to Palestine.

The groundwater aquifers are the primary source of drinking water. The largest of the three main aquifers is the Yarkon-Tanninim Aquifer. This receives water that has infiltrated from rainfall on the West Bank. The water moves west towards the coast. The

Yarkon-Tanninim aquifer supplies Israel with about 340x10^6 m^3 of water annually, largely to the Jerusalem-Tel-Aviv area[156]. (10^6 m^3 of water can typically supply a town of 20,000 people for a year, with each person using 150 litres per day). Israel's population is about 8 million, so the average water use per person per day is well in excess of 150 litres per day. Palestinians use about 20x10^6 m^3 a year for a population of 3.6 million, which is just under half of Israel's population. (Clearly, Palestinians use less water per person than Israelis). The next largest aquifer is the Nablus-Gilboa Aquifer. The catchment for this aquifer is on the West Bank, but most of the water emerges as springs in Israel. Israel takes about 115x10^6 m^3 a year, largely for agricultural irrigation in the kibbutzim (communes) and the moshavim (cooperative settlements) in Galilee. Lastly, the Eastern Aquifer supplies about 40x10^6 m^3 annually to the Israeli settlements in the Jordan Valley, and about 60x10^6 m^3 a year to the Palestinians. Some water is also drawn from the Sea of Galilee and the Jordan River for water supply and agricultural irrigation. The Jordan River is shared between Israel and Palestine both geographically and according to international law, though Israel is better placed to use it as a surface water resource.

The distribution of water between Israel and Palestine certainly needs revision. The World Bank (2009) has criticised the Israeli authorities for the unfair distribution of water to Israeli and Palestinian citizens[157]. It states that Israel has taken an increasing share of available water supplies, while making Palestine more dependent on Mekorot, the Israel water authority. However, Israel claims that the Palestinian economy has lagged severely behind the Israeli economy, and therefore has not needed more water. Israel also points out that domestic water use in Palestine increased by 640% between 1967 and 1995 due to fifty new wells drilled by Israel[158].

Israel therefore disputes the World Bank's view. Mekorot claims that Israel has never helped itself to water beneath Palestinian lands. Instead, Israel obtains 30% of its water from the Sea of Galilee and the Coastal aquifer, which are entirely within Israel's pre-1967 borders. Another 30% comes from the Western (83%) and North-eastern (80%) aquifers, with most of the water coming under the pre-1967 agreements on Israel's borders. In the 1950s, Israel used even larger proportions of these aquifers, so Israel claims that the Palestinian share has actually increased! In addition, Israel claims it supplies over 40x10⁶ m³ from its own resources, which it can barely afford, for Palestinian use on the West Bank, and including ten villages in South Lebanon, and an annual amount to Jordan.

However, the evidence is that the annual per capita usage of water by Israel in 1995 was 308 m³, while the corresponding Palestinian usage in the West Bank was 124 m³. In its defence Israel argues that it has made much more efficient use of water over the previous ten years when its population grew by 32% but its water use increased by just 3.3%; this was significantly less than either Jordan or Syria in the same period.

In 2008 there was a serious drought in the region. Due to chronic water shortage, per person consumption in the Palestinian West Bank dropped to 66% of the World Health Organization's recommended minimum daily amount of 100 litres per day, whereas the average daily water consumption in Israeli cities was 3.5 times higher. Obviously there are strong inequalities that reflect the current political realities.

The kingdom of Jordan lies to the east of the Jordan River. Running parallel to the river is the King Abdullah Canal, designed in 1957 and built in phases[159]. It was originally known as the East Ghor Main Canal. A corresponding canal was also planned for the West Bank of the river, but this was not built because Israel captured the West Bank from Jordan during the Six-Day War in 1967. In 1987 the existing canal was renamed after Abdullah I of Jordan. The canal has its source in the Yarmouk River, and receives additional water from wells in the Yarmouk valley and from wadis. The canal has a design capacity of 20 m³/s at the Yarmouk River, which reduces to 2.3 m³/s at its southern end. Water flows by gravity along its 110 km length, and is used to supply drinking water of 90x10⁶ m³ per year for Greater Amman irrigation as well as irrigation water along its length. This has allowed the cultivation of oranges, bananas, vegetables and sugar beet on the Jordanian side of the river. In contrast, Israel has constructed a water-supply grid pumping 320x10⁶ m³ (almost as much as the canal) per year from the Jordan River to the centre and south of Israel, which is very short of water.

There is little doubt that the water resources of the Jordan valley are under stress. It is important that all countries in the region work together to establish equitable arrangements for sharing water, and reducing the potential for conflict. The question is how this can be done most effectively.

We come across many long term disputes concerning water supported by well-established reasons on each side. The disputes are over water for irrigation and drinking, or for consumptive or non-consumptive use, or perhaps for navigation. Problems arise with the use and exploitation of water resources across national boundaries. International laws have been formulated, aimed at preventing conflict and encouraging cooperation of the shared resource. The chief international legal document is the United Nations Convention on the Law of the Non-Navigation Uses of International Watercourses[160] (UN General Assembly May 1997). This Convention applies to uses of international watercourses and their waters for purposes other than navigation, and to measures to protect, preserve and manage these watercourses and their waters. Water law has evolved through agreements between states, and contractual legal and technical arrangements. Most developed countries have water acts or directives in their legislation. For example, the European Union adopted the Water Framework Directive in 2000. The purpose of the Directive is to establish a framework for the protection of inland surface waters (rivers and lakes), transitional waters (estuaries), coastal waters and groundwater. It will aim to ensure that all aquatic ecosystems with regard to their water needs, terrestrial ecosystems and wetlands meet 'good status' by 2015.

While some users compete for the same water and struggle to increase their control over the resource, they also need to cooperate if they want to make effective use of water and sustain the water's quantity and quality in the long run. This often occurs in 'pluralistic' legal contexts, where formal and informal normative systems sometimes clash. For example, in South Africa, water management moved from a pre-colonial collective activity to a publicly regulated resource under Roman-Dutch law. It was then transformed under Anglo-Saxon jurisprudence when it was captured as a private resource to the benefit of a small minority. A main objective of the current 1998 Water Act in South Africa[161] is to redistribute water rights by granting access to water for all.

What does the future therefore hold for the use of water by Israel and Palestine? The organisation, Canadians for Justice and Peace in the Middle East (CJPME) has commented that[162]: 'under international law and as the occupying power, Israel holds the primary responsibility for the welfare of the Palestinians. Yet discriminatory Israeli policies and practices routinely deny Palestinians their right to water and sanitation. The right to water and sanitation, and the right for a people to make use of their natural wealth are human rights under international law that Israel has ratified[163]. Article 27 of the Fourth Geneva Convention prohibits an occupying state from discriminating between residents of occupied territory, yet Israeli colonists in the West Bank are allocated much more water than Palestinians.' In addition, the CJPME points out that 'Article 43 of The Hague Regulations (1907) prohibits the occupying power from changing laws that were in place prior to the occupation.' [164] Now that Palestine has been recognized by the United Nations as having observer status[165], there are good reasons for the international agreements between Israel and Palestine to be reviewed and redrafted.

References:

World Bank (2009) 'Assessment of Restrictions on Palestinian Water Sector Development' World Bank, New York

Burke III, E and Pomeranz, K (2009) The environment and world history. University of California Press

Resource questions:

How can we help Jews and Arabs in Israel and Palestine to work together?

In particular, how can the two peoples make best and fair use of their joint water resources?

37

Virtual water

> 'It takes 1,000 tons of water to produce 1 ton of grain. As water becomes scarce and countries are forced to divert irrigation water to cities and industry, they will import more grain. As they do so, water scarcity will be transmitted across national borders via the grain trade. Aquifer depletion is a largely invisible threat, but that does not make it any less real.'
>
> Lester A. Brown, Michael Renner, and Brian Halweil, in Vital Signs 1999, W. W. Norton, New York, 1999

Read Jeremiah 18:1-10

Fresh water is not only needed for human consumption but it is also vital for agriculture, food processing, and manufacturing and construction industries. Water is needed to grow staple crops, to rear animals, and to process foods; it is also used in the production of many household articles such as cleaning products, crockery, pots and pans, paper, furniture, televisions and washing machines as well as paints, glass and bricks and mortar! Indeed, most industries and manufacturing processes use water at some stage.

An interesting use of water in the Old Testament was in the production of clay pots. God deliberately sought to teach Jeremiah an important lesson through the local potter. God told him to go down to the potter's house; see Jeremiah 18:1-10. There Jeremiah observed the potter at work. The potter took lumps of clay and placed them on a horizontal wheel that he could turn, rotating the clay in his hands, which he then moulded into the particular shape he had in mind. Jeremiah watched the potter making a household pot. Despite the potter's skill, the pot became marred for some reason. There may have been too little clay in the lump he was using, or maybe the consistency of the clay was not right. Whatever the problem, by drawing on his long experience in making different pots, the potter decided to break down what he had been making and to make the pot to an alternative design. This he did successfully. God used this everyday example to speak to Jeremiah. He was teaching him that the LORD could fashion the conflicts and movements between nations as he wanted, and Israel was not exempt from being caught up in these changes.

However, this story reminds us that water is implicitly important in the pottery process. The lump of clay used by the potter consisted of fine particles of material, which would have been collected by the potter from a particular source. The potter would add water to the clay as necessary to make it more malleable in his hands on the wheel. Whatever the consistency of the clay, the potter could change it by adding more water, or by squeezing water out of the lump using his hands. Once the consistency was right, the potter proceeded to shape the pot. At this stage, because the pot still contained a lot of water, it could be marred because the form was not yet rigid—it could still be easily distorted. Most of the water had to be removed from the moulded pot. This was done by putting the pot to dry in the sun, or evaporating the water more rapidly from the pot by placing it in a hot furnace. As the water was being removed the pot did not simply disintegrate into its separate clay particles: the water produced a chemical reaction between the particles so that they continued to 'stick' together, retaining the shape formed by the potter. But the water had done its work. It had enabled the clay particles to come together in a flexible lump that could be moulded by the potter, and yet once that part of the process was complete, almost all of the water needed to be removed, leaving the pot with the desired rigid shape that would make it useful for a number of purposes—even containing water. Water therefore made an essential contribution to the development of the pot, but very little of it remained in the final product. This use of water also occurs in a large number of manufacturing processes in modern society. The availability of water for domestic use is important, but our industries today use far more water to make the material goods and products on which we are so dependent. Much of the fresh water that is used in these processes is 'consumed'; that is, the water may become highly polluted and either has its pollutants removed (which can be very expensive) or is disposed of as 'liquid waste'; some water evaporates into the atmosphere, or is trapped in the products.

The threat of significant drought in the first half of 2011 made national governments and water authorities in Western Europe anxious as they became ever more aware of the vulnerability of food resources and agriculture which depend critically on water. Estimates of the amount of water needed to produce staple foods such as grain, vegetables and fruit became important, especially as water used by plants either goes into building the plants' cellular structure or is lost to the atmosphere at some stage by evaporation; so it is not then available for other (processing or manufacturing) purposes. It is possible to determine how much water is used by different plants, how much is needed to raise animals for meat, milk and eggs, and how much is required for processing foods such as cheese and butter, jams and beverages. Indeed, we can do the same for most industrial products that we are so dependent on. We can then determine how much water is needed to produce these foods and products.

For example, it takes about 10 litres of water to produce 1 sheet of paper[166]. This includes growing the wood and processing it. About 40 litres of water are used in producing a loaf of bread, and 1,300 litres are needed to produce one kilogram of wheat. There are considerable variations in the amount of water needed to produce meat products. One of the largest amounts is 15,000 litres for one kilogram of beef, compared with 4,800 litres for one kilogram of pork. It takes about 70 litres of water to produce one apple, 120 litres for a glass of wine and 32 litres for a cup of tea! Clothes are also water hungry: a pair of jeans takes 10,855 litres of water while one kilogram of leather takes 16,600 litres.

The total volume of freshwater used to produce an item is called its water footprint[167]. This concept was introduced in 2002 by Aryen Hoekstra[168] at UNESCO-IHE. The idea of the water footprint is extended to define the total freshwater used to produce goods and services that are consumed by an individual or community, or generated by producers.

There are three components ('blue', 'green' and 'grey' water) that make up the calculation of the footprint. 'Blue' refers to surface and groundwater used to produce the appropriate goods and services, 'green' is the water evaporated from rainwater stored as moisture in the soil during the production processes, and 'grey' is the polluted water created.

As part of the idea behind the water footprint of an individual or community, we have the notion of a direct amount of freshwater used by the individual or community and an indirect or 'virtual' amount of freshwater needed to produce the products or services it consumes. We can then determine the water footprint for an individual in a given country. For the Netherlands, this is about 2300 m³/person/year, with 89% of the virtual water use coming from outside the country, whereas the global footprint is 1385 m³/person/year. An individual in China consumes about 7% of its water footprint from outside the country. For Israel in 2012, the footprint was 1391 m³/person/year with 74% coming from outside the country.

Some important implications follow from the figures above. First, enormous amounts of water are used in producing some products, beef and leather being two examples; we need to carefully consider whether such a use of water is justified. Second, we should aim to reuse water wherever possible. Any wastewater that can be collected from one process should be cleaned and made available for reuse, whether for the same process or for another. Some water may be lost, but cleaning and reusing makes better use of a scarce resource. Where possible 'grey' water should be treated and returned to 'blue' or preferably 'green' status. But there is almost always a cost involved. If the consumer is partly to bear this cost, then the prices of corresponding products may have to be increased. Only a small percentage of the water used in agriculture is available for re-use, due to evapotranspiration through plants or loss directly to the atmosphere by evaporation. Farmers in the Nile delta are so short of water for irrigation that they tend to reuse drainage water, even though the water can be heavily polluted with salts. Third, where water is scarce, it is better to focus production on low water-use products, while importing products with high water consumption from countries or regions where water is more plentiful. The fourth implication is that by importing and exporting products, countries are importing and exporting virtual water, that is, the water needed to make the product less the water that is available for recycling.

Note that the water footprint can be plotted geographically as an environmental parameter and it can vary in time. Also, it is only one parameter among several that will need to be considered when carrying out a proper analysis of the use of the water by individuals, groups and communities. For example, it may be important to have additional information on the local hydrology, climate, geology, topography, population and demographics. What is apparent is that the water footprint enables us to compare the need for water by different products, and therefore opens up ways of comparing products and services from different sources.

References:

Water Footprint Network. (2011) *Water Footprint Assessment Manual*, EarthScan

Global Water Footprint Standard.

http://www.waterfootprint.org/?page=files/GlobalWaterFootprint

Water Footprint Network: WaterStat.

http://www.waterfootprint.org/?page=cal/WaterFootprintCalculator

Resource questions:

Find out what it costs in terms of water per kilogram to manufacture a typical family car.

What products from which countries should you be concerned about with regard to its water footprint?

38

Water and natural disasters

'. . . there is no way to fully prevent flooding. Even if all preventive measures have been taken, the risk needs to remain firmly on people's agenda. We may have no choice but to live with floods, but suitable risk prevention in combination with appropriate insurance cover can significantly reduce their catastrophic effects.'
Dr.-Ing. Wolfgang Kron, Head of Research,
Hydrological Hazards in Geo Risks Research, MunichRE in
'Flooding – There is no such thing as complete protection'
from 'Topics Online', 05-07-2013, MunichRE

'By 2050, rising populations in flood-prone lands, climate change, deforestation, loss of wetlands and rising sea levels are expected to increase the number of people vulnerable to flood disaster to 2 billion.'
The United Nations World Water
Development Report 4 (2012) published by UNESCO

Read Psalm 104:1-18

Some of the most serious natural disasters in terms of human life have been due to floods, both from the ocean and in-land rivers, and from droughts. The media, including television and the Internet, make us immediately aware of such disasters. Graphic images of these disasters are on our screens within moments of their occurrence. We are becoming so familiar with the devastation and suffering experienced by those caught up in the floods or the emaciated human bodies that are a consequence of not having eaten for days that we are in danger of being inured to their horror and the devastating destruction of people's homes, communities and culture. We do not therefore, respond empathetically to the needs of those affected. The people who suffer most are those with few, if any, choices about where they can live, namely the poor and dispossessed. Often therefore, although floods and droughts can be regarded as 'natural' events in that their causes and generation are largely to be found in the continuing outworking of the original creation of the planet, such events raise issues of social justice. The events take on an increasing sense of importance for several reasons. One is that extreme flooding is becoming more frequent, ostensibly due to global warming. Then there is the savage intensity of droughts in particular regions where the population has been overexploiting its water resources. Also, the continental plates in the earth's crust are continually moving against each other, which introduce the sudden ground movements that can so devastate our attempts to order our environments on the surface of the planet. Many people in organisations dedicated to bring aid and resources to those affected by flooding are complaining that extreme flooding is becoming more frequent. There is a considerable body of evidence confirming this assertion, and many people are calling for governments to initiate further precautionary action to minimise the risk and damages.

The types of flooding event vary widely. Floods are made more acute by man-made changes to the catchments generating the runoff. Farmers often decrease the 'hydromorphological' resistance of the catchment, by draining wetlands, straightening streams and removing obstructions, while endeavouring to get the water as fast as possible to the next channel downstream. These farmers benefit while others downstream may suffer from an increased flood risk. Farmers also cut down unwanted vegetation such as trees to make more space for agricultural crops. This can mean that there is considerably less effective storage for flood water in the catchment, and much more erosion of fertile top soil. The loss of storage means that runoff is more immediate and larger, while the erosion of the top soil means there is a loss of fertility and increasing sedimentation in the channels downstream. Many developing countries have plundered their natural and tropical forests for the valuable timber, while being ignorant of the potential problems that the clearing of this valuable land cover introduces, namely, loss of habitat for rare endangered species, extensive erosion of rich soil accumulated over possibly thousands of years, and much higher rates of runoff of rainfall leading to extensive flooding in the rivers downstream.

Wherever we are in the world, it is very difficult for us to manage flooding, except in the most simple and straightforward of cases. We certainly cannot manage the weather, though we affect it by discharging carbon dioxide and other greenhouse gases into the atmosphere. We can tamper with the runoff of rainwater from natural and urban catchments, and hopefully direct flood waters (as excess rainfall-runoff) away from areas where they can cause damage into streams that have sufficient capacity to convey the excess water away safely. But our most effective way of managing floods is not by directly intervening in the floods, but by minimising our use of unprotected flood plains of rivers for our urban or transport infrastructures. The fact is that we are unable to resist the temptation to use such otherwise prime land: after all, it still provides the greatest potential for important and valuable communication corridors for our conurbations. Our city planners in particular have had to make compromises in using flood plain corridors, and in adopting these for building and construction. The Dutch made a profound choice centuries ago to live with this decision, with the result that the most valuable third of their land area in the Netherlands is below sea level and therefore subject to flooding risk from both rivers passing through the area and from high tides, surges and wave overtopping along their coast. Similarly, the people of the Maldives have continued to live in their paradise of islands despite the consequences of rising sea levels.

Ironically, other than the huge civil engineering works in North America and China involving the construction of massive dams on the Colorado, Yangtse and Yellow rivers, the other major river that does appear to have been tamed is the River Nile upstream of Egypt. The Aswan dam, with Lake Nasser behind it, has effectively provided Egypt with a guaranteed resource of at least 21 years of fresh water stored in the reservoir, and has all but eliminated the floods from the White and Blue Niles. This means that flood waters no longer inundate the agricultural land of the delta, to which they brought rich alluvial sediment to nourish the soil: the farmers now have to provide their own fertilizer and other materials to maintain the productivity of their land. But the control of flooding and the provision of the source of fresh water more than compensate for the associated disadvantages in the minds of the Egyptians.

The experiences the Children of Israel had of the regular flooding in the Nile in Egypt and in the Jordan, as well as the flash flooding events in desert places, probably raised considerable amounts of fear in their national psyche. To reassure themselves, they envisioned their God as ruling 'over the deep waters' (Psalm 29:10); in other words, God was in control of what they were unable to master and were inherently afraid of. The Bible has a number of references that affirm God's sovereignty over events in the natural world: see, for example, Psalms 24:2, 29:10, 32:6, 69:2, 95:3, 98:8. As we have seen repeatedly, water brings about life and death, construction and destruction. It was so important for the Israelites that God was in control of the events and phenomena they feared. It is no surprise therefore that some of the most dramatic examples of God's saving works involved water, with its potential to take away life. The crossing of the Red Sea, the crossing of the Jordan, the dramatic rain storm to end the drought in Israel brought about by Elijah's prayer, the stilling of the storm by Jesus on Lake Galilee, and the rescue of everybody on board the ship taking Paul to Rome, are all examples of God's interventions involving water.

God is good. He only has good gifts for his children, and plans for our good and not for evil (Jeremiah 29:11). We cannot accuse God of causing death through natural disasters, whatever form they take. God sends his rain on the just and the unjust. If we die because of it, then that is part of the physical processes in this universe. Whatever happens in our lives we are encouraged to express our on-going thankfulness to God for his grace, favour and salvation.

We live in a world where extreme random events occur which we cannot control. Sometimes we can forecast them and order our lives to minimize the risk of damage, but it is still appropriate to ask what we can do in response to natural disasters. How is it that as a consequence of these disasters many people suffer? Hurricane Katrina in New Orleans, SuperStorm Sandy on the eastern seaboard of the United States, the tsunamis in the Indian Ocean affecting Sumatra, the east coast of India, and Sri Lanka, and the north-west coast of Japan all resulted in enormous devastation; however, the devastation was in large part also due to the heavy density of population in these areas. If the storms or tsunamis had hit unoccupied land, the world's press would not have reacted as dramatically as they did[169]. Also, there was a clear sense of shock, a sense that these disasters were unexpected: these particular locations had not been affected by such disasters in recent human history. It is increasingly fashionable to find something or someone to blame when crises such as these occur, as if to make up for our ignorance or our neglect of the risks. We blame a more volatile climate or industrial development in other countries generating excessive amounts of carbon dioxide, or the destruction of the rain forests. Industry, shipping and air travel have raised the concentration of carbon dioxide in the atmosphere way above recorded levels in recent geological history, and the IPCC scientists are convinced that they are making a major contribution to global warming. If we build houses on the floodplains of rivers we must expect to get wet, at least occasionally. Because the earth is made up of a number of random processes with many interdependent parameters in which extremes of weather, volcanoes and earthquakes occur, and where there are catastrophes waiting to happen, we must expect to be at risk, especially when we find these events so difficult to forecast or predict. That is the nature of the world we live in. Jesus does not promise us safety from injury and death, but he confirms that with him we can achieve all that he calls us to do, despite the disasters that may happen around us or even to us.

This world is one in which sudden death happens, whether through accident or natural events in the environment. Fortunately we can reduce the impact of these events, but bad things do happen to good people. Jesus was asked whether the men who died when a tower collapsed on them were especially sinful. He replied that what happened to them was not a result of anything they had done or their state, but it happened so that God's glory could be revealed; see Luke 13:4. The fact is that our world and universe at the present time are subjected to frustration: they cannot be what God originally intended them to be until the story of humankind's redemption has been fully worked out. There is a conflict between good and evil, which has resulted in constraints on all parts of creation. Jesus came to resolve the situation by bringing about God's rule in this world, a rule that sets us free to become God's adopted sons through the indwelling Holy Spirit with all that that entails; see Romans 8:18-25. The same Holy Spirit, who was active with the Father and the Son in creation, yearns within us as we wait for the completion of our adoption as sons, which will be as much physical as it is spiritual. What is important is that we continue to thank God for all that he has done in our lives, and is going to do (Philippians 4:4-7), for we can have confidence that all things in the natural environment and in human society work together for good (Romans 8:28). At present the whole creation groans together as it waits for the consummation of history, when there will be a new heaven and a new earth; see Revelation 21:1.

Resource questions:

'. . . weeping may remain for a night, but rejoicing comes in the morning' (Psalm 30:5). The Bible is full of verses that express a relationship with God that is based on confidence (faith) that God is good and will vindicate those who put their faith and trust in him. How can we increase our faith in the light of so many natural disasters? As you recall your experiences of life, how would you explain to someone who has recently come to faith in Jesus that God will take care of them, no matter what happens in his or her life?

39

Water and the end of time

'Like a muddied spring or a polluted well is a righteous man who gives way to the wicked.'

Proverbs 25:26

Read Revelation 16

We have seen that water has played an important role in the creation and preservation of the earth. This chapter looks at the place of water at the conclusion of the earth's present existence. The Bible teaches that creation had a beginning, so it will have an end. But what do we mean by the 'end' of things? The Israelites in exile looked to their return to Jerusalem and the Promised Land as the culmination of their journey. The Old Testament 'major' prophets, such as (the second) Isaiah, Jeremiah, Ezekiel, and a group of 'minor' prophets including Amos, Joel and Zechariah, accepted that God would bring about the return of the exiles and the restoration of the kingdom of Israel. But from God's point of view it was critical that the house of Israel should realize that it was not for its own sake that the Sovereign LORD would do these things but for the sake of his holy name, God wanted to show his holiness to the nations through his people Israel. Then he would reverse the Exile: 'For I will take you out of the nations; I will gather you from all the countries and bring you back into your own land. I will sprinkle clean water on you, and you will be clean; I will cleanse you from all your impurities and from all your idols.' (Ezekiel 36:24-25).

The early Christian Church had a different interpretation of the Old Testament scriptures, which they augmented with other writings, in particular, the book of Revelation, written by the Apostle John. While in exile on the Isle of Patmos, John had a series of visions which formed his view of the Apocalypse. (This concerns the disclosure of knowledge, or something hidden. Today the word is commonly used to refer to the end of the world). In the book of Revelation, specific references to water in one form or another describe the catastrophes which will befall the human race and the whole of creation as the titanic battle between God and satan is finally resolved. Just as water was the source out of which all things were created, so water appears as the symbol of eternal life for all who will come and drink of the living springs of the Spirit that flow in complete abundance from God.

In John's description of the approach of the end of all things, God sends angels to bring about a series of seven plagues. These parallel the plagues which God brought about through Moses when he confronted Pharaoh prior to the Exodus from Egypt. For 'Pharaoh' now read 'satan'. Water in some form or other appears in the first three plagues. John says that the first angel sounded his trumpet and a mixture of hail, fire and blood was hurled down upon the earth. When the second angel blew his trumpet a blazing mountain was thrown into the sea; a third of the sea was turned into blood, a third of the living creatures in the sea died, and a third of all ships were destroyed. Then the third angel sounded his trumpet and 'A great star fell from heaven, blazing like a torch, and it fell on a third of the rivers and on the fountains of water. The name of the star is Wormwood. A third of the waters became wormwood, and many people died of the water because it was made bitter' (Revelation 8:6-11). Water in the environment is polluted and makes life critically difficult for all human beings.

After six angels had sounded their trumpets, God gave power to two witnesses who, like Elijah, could stop the rain, and like Moses, could turn the waters into blood, to strike the earth with any kind of plague as often as they wanted. The beast that came up from the Abyss, however, attacked and killed them. Their bodies lay in the street of the great city (symbolizing Sodom, Egypt and Jerusalem) for three and a half days. Then they came alive again through the breath of God, who took them into heaven.

After the seventh angel sounded his trumpet, a pregnant woman gave birth to a child which was under immediate threat from the dragon. War followed between Michael and his angels and the dragon and his angels. The dragon targeted the woman and her son on the earth, attempting to drown them in a river of water which came from his mouth. Again, the death inducing feature of water is used. But the earth swallowed the water, and the woman escaped out of the dragon's reach. Subsequently two beasts came out of the sea and from the earth to deceive people. John writes: 'Then I saw another angel flying in midair, and he had the eternal gospel to proclaim to those who live on the earth—to every nation, tribe, language and people. He said in a loud voice, 'Fear God and give him glory, because the hour of his judgment has come. Worship him who made the heavens, the earth, the sea and the springs of water.' (Revelation 14:6). The battle for the hearts and minds of human beings was at its peak.

After this God announced it was time to 'harvest' the earth, and the seven angels were told to pour out the seven bowls of God's wrath onto the earth. The first bowl caused ugly and painful sores to break out on people who had the mark of the beast. The second bowl was poured out on the sea turning it into blood (as with the breaking of the second seal), and everything living in it died. The third angel poured his bowl on the rivers and springs of water and they became blood, thus making all sources of fresh water unusable. The other angels poured out the remaining items of God's wrath including: 'The sixth angel poured his bowl on the great river Euphrates and its water was dried up'. This enabled the kings from the east to gather across the boundary formed by the River Euphrates with the kings from the whole earth for the great battle at Armageddon (Revelation 16:12). The climax came when the seventh angel generated earthquakes and caused huge hailstones to fall from the sky. It was then that the 'great prostitute', 'Babylon the Great' sitting on 'many waters' and drunk on the blood of the saints, was revealed to the Apostle John on Patmos. The 'waters' in this case are peoples, multitudes, nations and languages under the influence of Babylon. The kings associated with the beast brought Babylon to ruin through their hatred of her. The merchants, who had grown rich through international trade facilitated by the transport of goods by sea, saw the economic collapse and ruin of Babylon and lamented her demise because she had fostered the demand for their goods and therefore been the main reason for their prosperity. This is very much like the world economy today, in which goods are transported by sea from one country to another in response to existing demand in the developed countries and growing demand in the BRIC (Brazil, Russia, India and China) and other developing countries. Although goods are also transported by air, and the service sector is becoming increasingly important, shipping is still at the heart of international trade. Revelation talks about the demise of the whole structure of international trade as we know it. Not that there is anything intrinsically wrong or evil about such trade; indeed Isaiah talks about the way in which the Exile will come to an end when the wealth of the nations is brought to Jerusalem over the seas (Isaiah 60.5). What is wrong is the attitude of people to such trade, where the 'love of money' is at the root of their approach to making wealth along with a ready acceptance of corruption to facilitate it.

Back to Revelation! At long last, a horse and its rider appeared. The rider called Faithful and True, namely Jesus Christ the Lamb, LORD of LORDs and King of Kings, came with justice to judge and to make war against the beast. He overcame the dragon, also named the ancient serpent, the devil or satan, and threw him into the abyss for a thousand years (that is, what is popularly known as the Millennium[170]). At the end of that time, satan was released to attack the city of God, which he subsequently surrounded. But fire came down from heaven to destroy satan and his army. He and all his forces were finally thrown into the lake of burning sulphur, where they are tormented forever. Then came the judgment of all people and things when Jesus was revealed in his glory.

This brief overview of the events in Revelation makes up what are called 'the last times'; the events describe a sequence of ecological catastrophes. The blessings of water given for all human beings are withdrawn; indeed all life suffers as a consequence. The state of darkness and the waters of 'chaos', which existed before creation, return as an indication of the reversal of creation. It is from the deepest parts of the ocean, which is seen as a major source of evil that the beast emerges. This hideous being slanders God, destroys humankind and brings destruction on the whole earth (Revelation 13:1).

The history of Israel in biblical and modern times has involved considerable dislocation from the land and intense periods of suffering. It was Assyria that attacked and dispersed Israel: 'See how the waters are rising in the north; they will become an overflowing torrent. They will overflow the land and everything in it, the towns and those who live in them. The people will cry out; all who dwell in the land will wail at the sound of the hoofs of galloping steeds, at the noise of enemy chariots and the rumble of their wheels. Fathers will not turn to help their children; their hands will hang limp.' (Jeremiah 47:2). The people of Judah were later taken into exile in Babylon.

Subsequently, it was only slowly that the Jews returned to Jerusalem and rebuilt the temple. But for those who returned life was difficult, especially as they wanted to maintain the purity of their race. Babylon was destroyed; Alexander the Great dominated the Middle East for a brief period before the Romans brought an uneasy stability to the region through a ruthless, imposed control. The emergence of Christianity, and then six centuries later the rise of Islam, brought a different sort of conflict for Judaism. For more than one and a half millennia there has been an uneasy peace between the three major religions, all of which look back to Abraham as their ancestor and have claims on Jerusalem. The fluctuating fortunes of Jerusalem between the three religions, particularly since the eighth century, are symptomatic of the relationships between them. In the 20th century during the holocaust, the Jews suffered horribly in Europe at the hands of an ostensibly Christian nation. The re-establishment of the Jews in modern day Israel in 1948 and the subsequent volatility of the region have given rise to considerable uncertainty, with wars between Israel and Egypt, and Iran and Iraq, and civil wars in Libya and Syria to name but a few. Arab and Jew have found it difficult to coexist, and American support for Israel distorts the balance of power in the region. It is small wonder that some extremists within Islam and Christianity particularly seek to provoke Armageddon sooner rather than later. However, it is God who determines the timetable of events.

For further information read the following:

Wright, T N (2011) *Simply Jesus.* SPCK

Resource questions:

What do you think about the establishment of Israel as a nation and the declaration by the United Nations in 2013 that Palestine is accorded non-member observer status in the general assembly?

Do you think there is evidence that we are approaching the 'end time'? What evidence would you put forward for this?

Are you expecting the second coming of Jesus very soon?

40

Water in paradise

'*Before me green slopes made a wide amphitheatre, enclosing a frothy and pulsating lake into which, over many-coloured rocks, a waterfall was pouring. Here once again I realized that something had happened to my senses so that they were now receiving impressions that would normally exceed their capacity. On Earth, such a waterfall would not have been perceived at all as a whole: it was too big. Its sound would have been a terror in the woods for twenty miles. Here after the first shock my sensibility took both as a well-built ship takes a huge wave. I exulted. The noise, though gigantic, was like giants' laughter: like the revelry of a whole college of giants' together laughing, dancing, singing, roaring at their high works.*'

C. S. Lewis, (1946) The Great Divorce, Collins

'*The awakened take on the quality of water, which is soft and pliant yet irresistible in its power, which does not strive, yet benefits all beings. By their egoless action others are transformed through their detachment the whole world prospers; owing to their desirelessness others are left unspoiled. Water is drawn out of the river to irrigate fields. The water is quite indifferent to whether it is present in the river or in the fields. Thus it is that the enlightened act and live sweetly and powerfully in accordance with their destiny.*'

'*The heart of the enlightened' by Anthony de Mello*

'*He said to me: 'It is done. I am the Alpha and the Omega, the Beginning and the End. To him who is thirsty I will give to drink without cost from the spring of the water of life.*'

Revelation 21:6

227

> *'The deeper the waters are, the more still they run'.*
> *Korean proverb*

Read Revelation 22

With the final defeat of satan, a new heaven and earth will come into being. In particular, the physical earth and the spiritual heaven will be reintegrated, just as they were originally in the Garden of Eden before the Fall. The reality of this reintegration was made apparent in the risen Christ, who was the first fruits of the new creation. Although present in another world, he was able to be present also in our physical world: he could be touched and he could eat fish, yet he would come and go at will. The Bible makes it clear that the new heaven and earth will have some remarkable features. For example, the mountains and hills of Israel will come alive. 'In that day the mountains will drip new wine, and the hills will flow with milk; all the ravines of Judah will run with water. A fountain will flow out of the LORD's house and will water the valley of acacias.' (Joel 3:18). In Revelation it is stated there is no sea (such as the Mediterranean Sea, though apparently not including the Dead Sea) from which satan or the beast can emerge, in stark contrast to the original creation in which the sea is a key part of the hydrological cycle (Genesis 1:9-10) as well as a resting place for many who had died (Revelation 20:13). In the new heaven and earth there can be no hydrological cycle as we know it, with no sun (Revelation 22:5) providing the energy to drive it. Therefore we must assume that God provides the input directly to whatever hydrological cycle there will be in the new creation. We can probably also assume that the physics of the new creation may be similar but different to our present universe. Certainly, the relationships between God and human beings will be different than before: 'For the Lamb at the center of the throne will be their shepherd; he will lead them to springs of living water. And God will wipe away every tear from their eyes.' (Revelation 7:17). The Jewish and Israelite exiles will be brought back from lands where they were domiciled. The LORD says: 'They will come with weeping; they will pray as I bring them back. I will lead them beside streams of water on a level path where they will not stumble, because I am Israel's father, and Ephraim is my firstborn son.' (Jeremiah 31:9). 'They will come and shout for joy on the heights of Zion; they will rejoice in the bounty of the LORD—the grain, the new wine and the oil, the young of the flocks and herds. They will be like a well-watered garden, and they will sorrow no more.'(Jeremiah 31:12)

So far as Zechariah is concerned, the day of the LORD will be distinctive. 'On that day there will be no light, no cold or frost. It will be a unique day, without daytime or nighttime—a day known to the LORD. When evening comes, there will be light" (Zechariah 14:6).

The river flowing from the throne of God

Following Walton, does this mean there will be no time?[171] On that day living water will flow out from Jerusalem, half to the eastern sea and half to the western sea, in summer and in winter. Ezekiel is in conflict here with John in Revelation who says there will be no sea. The LORD will be king over the whole earth. 'On that day there will be one LORD, and his name the only name.' (Zechariah 14:6-8). Together with Ezekiel and Revelation, Zechariah makes much of the fact that a river of fresh water will flow from the throne of God established in Jerusalem. The river from the throne will sustain all life and will be made available to all the earth. Just as the rivers flowed from the Garden of Eden to water the earth and to sustain its life so the river from the throne of God will dispense God's life, healing and sustaining power and grace wherever it flows.

How does the geography of the area including Jerusalem affect the biblical accounts of the new heaven and the new earth? Jerusalem is about thirty-four miles (fifty-five km) east of the Mediterranean coast and surmounts a geological outcrop called the Ophel-Moriah ridge. The ridge is about 2,500 feet (762 m) above sea level. The northern end of the Dead Sea is a further twenty-four miles (thirty-nine km) east of the city. There are valleys on three sides of the city: the Hinnom Valley to the west and south, and the Kidron Valley to the east. There are no rivers or perennial streams flowing through or round the city, but there are a number of springs as well as a sophisticated system of cisterns, reservoirs, conduits, and aqueducts; see Chapter 16.

Ezekiel however, envisages a river flowing from the sanctuary in the temple into the Kidron Valley and away to the south-east. The actual route depends on the topography (which may well have changed down the centuries). Given a digital terrain model of the area, such as could be obtained today using LIDAR, for example, then it is a straightforward task to identify the potential channel network that describes runoff from rainfall over the area, or, given a particular starting point, the subsequent route of the channel from the source. Needless to say, as water flows downhill under the influence of gravity, the channel will follow the line of steepest descent. This would be subject to any geological or other change to the surface terrain. The source of the river is the sanctuary. This symbolizes the flow of forgiveness and salvation from God, the only source of life to a world that desperately needs it, and therefore we can assume that God generates the river.

In Ezekiel 47, the prophet is invited by the angel to measure the depth of the flow in the river from the sanctuary at a thousand stadia intervals stadia (where a stade is typically 600 ft or 183 m). The first measurement is ankle deep, enough to be refreshing on the feet, possibly with the sound of moving sand or gravel if the velocity of the flow is large enough. One thousand downstream the flow is knee-deep. This is becoming significant. We are not told what the speed of the flow is. If it was large enough then standing in water up to the knees would be sufficient to cause a person to lose his or her balance. Certainly a car in such a depth of flowing water could float and be carried away. (This should be born in mind when trying to drive through a flood!) Going another thousand stadia the river is flowing chest deep. Here we are told that a person is unable to stand; instead, being caught in the flow he or she would be carried downstream. There are some unexplained features about the river. For example, the implication of the description is that the discharge (the volume across a channel section per unit time) increases as you go down stream, unless the slope if the river bed gets flatter. If the discharge does not increase then in normal circumstances the channel would have a smaller width downstream and the bed gradient of the channel would decrease. This does not appear to be the case with the river in Ezekiel and Revelation, in which the discharge patently increases downstream. One way this could happen is for water to be contributed to the river from groundwater along the river. But the authors appear to imply that all the water in the river originates from the throne of God. Of course, this is in the new heaven and the new earth, which means that all sorts of other possibilities emerge. For example, the water could be generated in the river, much like Jesus breaking the five loaves and two fish so that every one of the five thousand men (plus women and children) had enough to eat. (Assuming that all the water comes from the throne of God, I prefer the multiplication solution.)

In John's vision, the river of the water of life, which is clear as crystal, flowed down the middle of the great street of the new Jerusalem from the throne of God and the Lamb (Revelation 22:1-2). Associated with the river is the tree of life, which Adam and Eve were prevented from approaching by an angel with a flaming sword. This tree bears twelve crops of fruit, yielding a separate crop each month. The number twelve is symbolic of the world-wide church as it was of the twelve tribes if Israel; none of the twelve months is without a specific fruit. The fruit is to be used as food, whereas the leaves of the tree of life are for the healing of the nations. This is similar to the vision of Ezekiel. Ezekiel saw a grove of trees on both banks of the river (Ezekiel 47:7) that also bears fruit every month (Ezekiel 47:12). Again, the fruit of the trees is good for food, and the leaves of the trees have healing properties. Thus both properties of the tree of life in Gen 3:22, namely the fruit for food and the benefit of having eternal life (the healing), are consistent with John and Ezekiel's visions.

Flowing away from Jerusalem the river irrigates the tree of life or the grove of trees along its banks that produces fruit every month (Ezekiel 47:12). Thus the life that flows from the throne of God transforms everything and anyone who will receive it. The physical water of the river is like the blood of the Messiah from the cross of Calvary that began as a trickle (John 19:34) and becomes a flood of redemption for all people (Revelation 1:5). So the flow from the cross benefits all, including Israel (see Zechariah 13:1-6; Revelation 1:5-6). Jesus told the woman at the well that the healing waters of God associated with the Word of God would be a 'spring of water welling up to eternal life' internal to each believer; see John 4:14. Likewise, in John 7:37-38 streams of living water are presented as flowing from the body of believers, suggesting the life-giving flow that comes into a person's life through faith in God and Jesus Christ.

The tree of life bears fruit for food and leaves 'for the healing of the nations' (Revelation 22:2), that is, the nations continue to have an integral place in heaven but they need 'healing', presumably meaning protection from rather than correction of any illness. The important thing is that humankind now has full access to the tree of life; see Gen 3:22-24. The healing of the nations could be during the proposed 'millennium', the thousand years before Jesus finally comes to destroy satan. In this period the Church will have the responsibility to give the gospel to all the nations and so bring about the healing of their spiritual state. This, of course, will not be needed in the final and perfect new Jerusalem.

Ezekiel has a fuller description of the impact of the river of life. The river teems with fish waiting to be caught, and contributes to life wherever it flows. In particular, it flows from Jerusalem down to the Jordan River and on to the Dead Sea. The water of the Dead Sea has approximately 25% minerals by weight compared with about 5% minerals of normal sea water. This is highly toxic to fish and therefore there are none in the Dead Sea. On reaching the Sea, the water from God amazingly causes the Dead Sea water to become fresh, that is, the salts are removed, possibly by coagulation, and the water is full of numerous species of (fresh water) fish; the unhealthy are healed and the dead become alive (Luke 7:21-22). However, salt water continues to remain among the reeds at the coast of the Sea, possibly to provide a source of salt for domestic use.

The early church spiritualized the visions of both Ezekiel and John. For example, the increasing flow in the river represented the growing number of believers. The fourfold measurement of the depth of flow represented the four Gospels with the fourth and deepest measurement identified with the Gospel of John, which was considered the most profound. The river was associated with baptism. Such interpretations can add value to the spiritual perceptions of Christians, but it can be illuminating to interpret these passages in terms of the literal as well as the symbolic elements.

What Ezekiel and John saw in their visions were real rivers with water that had life-giving and healing properties flowing from God's throne to human kind. For Ezekiel the fountain in the temple generated water that went as a river east to the (Dead) Sea. It was the source of blessing from the LORD to all. There will be no more curse on the land, the waters, on humans or animals; the desert will blossom like the rose; and the barren land and the bitter waters will be healed. The final abode for humanity is still on (the new) earth. God comes down and dwells with humans, bringing to earth the attributes of heaven.

Each person will see the divine glory revealed in the face of Jesus Christ. They will be like him when they see him and they will be known by him. Their sonship will be known to all; there will be no check on the free flow of mutual love between the members of Christ's family.

Finally, there is no darkness; there is no hidden corner where the light which comes from God cannot reach. The Bible begins with water and darkness, and ends with water and light! In between we are made aware of the overall importance of Jesus Christ, the only Son of God, who is the light of the world. But more than that: Jesus said of himself that he is the Way, the Truth and the Light; no-one comes to the Father but by him; see John 14:6. Finding the way through water therefore resolves into finding the way through Jesus Christ. May God give us grace to do so.

Reference:

C S Lewis (1946) *The Great Divorce*. Harper SanFrancisco

Resource Questions:

Does the vision that John describes impress you? Give reasons for your answer.

What are you especially looking forward to in the new heaven and earth?

Conclusion

Water acts

> 'For those who have the luxury of clean water running from the tap, the signs of water crisis are not apparent enough to change behaviour. But if we wait until those signs are apparent to all, it will be too late. That's why we must build awareness to include economic, social, humanitarian and national security indicators—and do it now.'
>
> Kristin Rechberger, Vice President, Corporate Partnerships, Mission Programs at National Geographic Society in 'Watermanegement – A Global Innovation Outlook Report', IBM, 2009

Read Ezekiel 47:6b-12

Fresh water is an increasingly scarce global resource, not because there is less fresh water year on year, but because of the growing demand of the world's population. Fortunately, many people have access to and the use of fresh water at the local level; in addition, they benefit virtually from the water resources of others when such resources are used to provide goods and services from a distance. Therefore when thinking about the sustainability of our water resources, we need to think globally and act locally, as suggested by the Scots town planner and social activist, Patrick Geddes[172]. Even if we are fortunate enough to have large resources of fresh water locally, we still have obligations and responsibilities to ensure that we make good use of these resources, protecting them from pollution, and using them sensibly to benefit others as well as ourselves. We need to be more aware of the vulnerability of people living downstream in a river basin; we can all become more conscious of the hidden water used in producing goods and services that we acquire from others, water which the producers may not be able to afford. It is imperative that social justice should be promoted in the management and distribution of water resources at whatever scale.

We are not short of water on our planet; more than 70% of the earth's surface is covered by water or ice. The problem is that most of this water is toxic when we drink it, due to dissolved salts. Nevertheless, the oceans are a critical source of water vapour that evaporates from the ocean surface into the atmosphere due to the radiation from our sun. In this way the sun drives the heat engine, which is responsible for our weather and climate. In particular, water vapour in the atmosphere is the principal greenhouse gas that affects climate change. It is also the source of precipitation onto the earth's surface that is essential in providing us with fresh water. So we are limited by the amount of fresh water available to us: we can barely afford the energy it takes to remove the salts from sea water artificially to increase our supply of fresh water.

Our bodies are about 70% water. Therefore water is critical to sustain our lives. What is more, we share water with each other, as well as with all other animal and plant life on the planet. We cannot avoid the implication that water makes us interdependent with all life. If a water molecule had expectations, it could expect to move successively from, say, a human body to a drain, a wastewater treatment works, a river, a fish, a stream, a groundwater aquifer, a spring, a plant, the atmosphere, a cloud, an ice crystal, a raindrop, the ocean surface, the deep ocean, and so on. There is no other substance on earth that can go through so many potentially different situations, and be used for so many different tasks and functions. This should fill us with awe. Apart from the implications for our physical world, water symbolises the supply of life-giving resources and the disposal of unwanted waste, sustenance and cleansing, birth and death, disaster and recreation, horror and beauty. As Christians we use water for baptism with the rich symbolism of being buried in the tomb with Christ in his death when immersed (or sprinkled) in water, and being raised with Christ to new life when coming out of the water. Water is essential for the growth and protection of the baby in its mother's womb as well as being an important symbol of the new birth in the Holy Spirit. Water that has been blessed is sprinkled on the congregation by a priest to signify cleansing, and maybe used by individuals when they make the sign of the cross. A number of churches also mix a small amount of water with wine in the Eucharistic chalice to symbolise the bringing together of the human and divine in the body of Christ. Because water plays an important part in our physical health and wellbeing, as promoted, for example, by the many 'spas' around the world, it becomes an important symbol both for physical and spiritual health, protection and cleansing.

There is now little doubt that present and future generations face enormous problems in managing and adjusting the earth's resources (i) in the light of our gross over consumption of non-renewable resources, (ii) in adapting to the accelerating growth in the world's population, and (iii) in limiting any unacceptable increase in global warming due to our industrial and economic development. The problems are so serious that we should ask ourselves if we are in danger of bidding farewell to this world as we know it because of our selfish and unthinking approach to life. Can we continue to neglect God's gifts of fresh air, fresh water and soil uncontaminated by industrial waste?

It is God's will that we care for this world, that we respect the limits that he has set, that we cherish the gifts which make life on planet Earth possible, and that we share the gifts with all who are in need—whether human beings, animals or plants. We need to have our values, beliefs and attitudes transformed into those of God's kingdom. We are responsible for working out our approaches to the problems of our world with the creative and sustaining insights of the triune God.

For example, we need to recover and maintain the pristine quality of groundwater, surface water in streams and rivers, and the oceans. We can do this by using less aggressive chemicals and detergents in our homes, consciously buying those products which are not poisonous, and which can decompose in the environment without disastrous effects. Industry needs to avoid releasing any potentially dangerous chemical into the environment, providing treatment for wastewater at source and making full use of wetlands for the natural extraction of pollutants. Those organisations we have mandated to look after our water resources should do so with the highest standards of care for the environment and using the most appropriate technologies.

The demands of agriculture in the past, such as increasing production of a particular food crop (for example, bananas or pineapples) to meet commercial demand while limiting the growing of other crops needed by the local population, or of a range of crops to meet the needs of a besieged population during wartime (for example, Britain during the second World War), can introduce changes to the land drainage that cannot be maintained in terms of social justice or are not required in peacetime. The temptation has been to drain lands that may be termed 'wetlands' to take advantage of their potential fertility for agriculture. Such drainage usually involves cleaning and straightening streams to improve the flow of water from the wetland. This 'improvement' limits the storage of water in the catchment, creating faster runoff with a larger peak discharge, causing probable adverse consequences for flows downstream. The 'improved' drainage decreases the hydromorphological resistance of the catchment. Where possible, such resistance should be restored by encouraging the natural development of the stream network.

We should aim to preserve natural aquatic habitats of fish, amphibians and many other animals, and of plants, for which living in water is important for their survival and the maintenance of the pool of their species. Often the water ecosystems around us are unique and very sensitive to change. Therefore these habitats need protection from pollution by the irresponsible behaviour of people in our communities or the careless attitudes of our authorities or industry.

Where there is the potential for drought we should use water sparingly, whether in the house or in the garden. It is better to have a short shower than a bath. Where possible we should use toilets with a dual flush or reduce the volume stored for each flush. We can consider having our cars steam washed, or harvesting rainfall from our house roofs or from garden sheds. We should consider our use of water seriously, not only directly but virtually. We can take an interest in where the things we buy come from. We can work out approximately how much water was used in manufacturing, and whether the cost in water is too much for the country of origin, especially if that country faces water shortages. We can evaluate the water footprint of an organization or an individual and decide whether to buy or not.

We can take a closer interest in how our water service providers make decisions, especially in reducing leakages from distribution networks and minimising pollution from sewerage overflows discharging to rivers, lakes or coastal waters. We should ask questions about the sustainability of water resources and the efficiency of asset management. We can encourage city authorities to increase infiltration of water into the ground in urban areas, using pervious pavements in car parks and infiltration trenches. We can request urban planners to give more attention to overland flow paths for flood water in urban areas and making buildings and infrastructure more flood resilient. We can do our own research into how people in areas of water stress, especially in developing countries, manage their water resources, and seek to support them in achieving their ambitions and goals. Where possible, we should support and care for the aquatic environment, especially the restoration of rivers and catchments to their natural (but managed) state. We should encourage and support the existence of non-government organisations (NGOs) that monitor water quantity and quality in the environment and contribute to the local government's development of water resource standards.

There may be many ways to improve access to drinking water for people and communities in developing countries, and to provide better opportunities for people depending on water, such as fishermen and farmers. We can introduce the best water—saving practices in our churches and inform the public about them. We can also support or develop campaigns informing society about water issues in our area and about best possible practices related to water.

Because of the increasing environmental, economic and social costs connected with drinking water distribution and delivery, water treatment and sanitation, we should aim to develop and support local and regional solutions for water systems conveying clean drinking water requiring minimum treatment and by the shortest possible distribution route from local resources to the users. At the same time, used (waste)water should be returned to the environment purified if possible at source.

There is little doubt that of all substances in our created universe, water is one of the most important from a physical and a spiritual point of view. Everything we are and do seems to depend in some way on this weird substance we call water. Perhaps in reading this book and doing your own research into different related topics you have gained new insights, identified other issues to do with water than are addressed here, and linked different aspects of human society, culture and religion with water. I would gladly know your thoughts and insights. For my part however, I would think that my task is done if I have encouraged you to continue championing water as a strategic, physical and spiritual resource for all humankind.

Actions: How about drawing up a proposal you could put to your church council about how your church community could address the problems and opportunities to do with water in your local area and further afield, especially in developing counties, and work more closely with public and private organisations involved in the water sector?

End Notes

Preface

[1] For a well written piece on the properties and nature of water see Bryson, 'A short history of nearly everything', Chap 18, p330

[2] See the 'DamBusters: the race to smash the dams 1943' by James Holland

Chapter 1: Water Crisis?

[3] Eskimos have 30 different words for snow

[4] Read the United Nations World Water Development Report 4, which is planned to be published every two years. The latest was produced in 2012

[5] Minimum amount of water needed by an individual per day as recommended by UNDP

[6] See the site for Institute Water for Africa at http://www.water-for-africa.org/en/water-consumption/articles/water-consumption-in-africa.html

[7] The United Nations World Water Development Report 3: p31

[8] World Water Development Report 4. World Water Assessment Programme (WWAP), March 2012.

[9] Human Development Report 2006. UNDP, 2006

[10] 50 l of water a day are needed by an individual person

Chapter 2: Water on Earth

[11] See the Wikipedia article on 'Water Distribution on Earth' (accessed on 05-11-2013).

[12] The actual height to which the oceans would rise is very difficult to determine, not least because the rise depends on the average surface temperature of the planet, the expansion of the oceans water as they heat up, the volume of water represented by the melting ice in the glaciers, over Greenland and Antarctica

[13] El Niño. Chinese climate scientists, among others, have successfully correlated the strength of El Niño events with trends in China's weather over periods of several months.

[14] The major nuclear disaster at Chernobyl in 1986 generated a large plume of radioactive fallout over western Russia and Europe. There have been many consequences of the fallout for the environment and human society. For example, the United Kingdom was forced to restrict the movements of more than four million sheep between 10,000 farms in order to protect the human food chain from contaminated meat.

Chapter 3: Water and Creation I

[15] See a translation of the text of Enuma Elish at http://www.ancienttexts. org/library/mesopotamian/enuma.html (accessed on 05-11-2013).

[16] Walton, J (2009) The Lost World of Genesis One. Proposition 3, p 36.

[17] Ibid p 18

[18] Ibid p 18

[19] Ibid p 18

[20] Ibid p 43

[21] Petersen E (2005) 'Creation and the gift of time', in 'Caring for Creation', ed S Tillett, Bible Reading Fellowship

[22] Ibid p 55

[23] Calvin J (1554) 'Commentary on Genesis Christian Classics Ethereal Library', Grand Rapids MI. 'For it appears opposed to common sense, and quite incredible, that there should be waters above the heaven. Hence some resort to allegory, and philosophize concerning angels; but quite beside the purpose. For, to my mind, this is a certain principle, that nothing is here treated of but the visible form of the world. He who would learn astronomy, and other recondite arts, let him go elsewhere'.

[24] Walton, ibid p 56

[25] Ibid p 58

[26] Ibid, Proposition 6, p 62

[27] Drane J (1990) 'An introduction to the Bible'. Lion Publishing. p250

[28] Islam states that God made man out of water (Qur'an 25:54)

[29] Consider the philosophic and theological arguments posed by Christians such as Francis Schaefer, John Polkinghorne and Tim Keller

[30] Petersen, ibid p22

[31] Walton ibid p 101 Proposition 11.

Chapter 4: Water and Creation II

[32] Two of the four rivers are easy to identify, whereas the other two are not. It is another area for scholars to debate. Perhaps the most likely solution is to be found in rivers that are actually near to the Tigris and the Euphrates.

[33] See Marks, J H (1971) The book of Genesis in The Interpreters Commentary on the Bible. SPCK

[34] See Walton on stories of creation, ibid p 107 and following

35 John Polkinghorne in his book Science and Christian Belief, SPCK, 1994, draws attention to the work of Ilya Prigogine for an argument that life is becoming more complex

36 Ilya Prigogine. For a concise description of his life and work see the corresponding Wikipedia article (accessed on 05-11-2013)

37 Stephen Hawking: 'We have made good progress on the first part, and now have the knowledge of the laws of evolution in all but the most extreme conditions. But until recently, we have had little idea about the initial conditions for the universe. However, this division into laws of evolution and initial conditions depends on time and space being separate and distinct. Under extreme conditions, general relativity and quantum theory allow time to behave like another dimension of space. This removes the distinction between time and space, and means the laws of evolution can also determine the initial state. The universe can spontaneously create itself out of nothing.

Moreover, we can calculate a probability that the universe was created in different states. These predictions are in excellent agreement with observations by the WMAP satellite of the cosmic microwave background, which is an imprint of the very early universe. We think we have solved the mystery of creation.' This is taken from the transcript of a talk given in 2008.

38 Sam Berry explains his argument in 'Caring for Creation', ed S Tillett, Bible Reading Fellowship, p 39

Chapter 5: Water and the Fall

Chapter 6: Water and Climate Change

39 Plimer, I (2009) 'Heaven and Earth: Global warming, the missing science'. Taylor Trade

40 Ibid p48

41 Ibid p49

42 Ibid p50

43 Ibid p52

44 Ibid p53

45 Ibid p53

46 Ibid p57

47 Ibid p59

48 Ibid p61

[49] Ibid p63

[50] Ibid p79-p85

[51] Ibid p435-p493

[52] IPCC 2007

[53] IPCC 2013

[54] See the article titled 'Global warming hiatus puts climate change scientists on the spot' in Los Angeles Times 22 September 2013

[55] IPCC 2013 ibid

[56] The Tambora explosion created a crater approximately 6 km across and 1 km deep. Other famous volcanoes that erupted in recent times include the volcano in the eastern Mediterranean that exploded in 365AD. It caused significant uplifts of the earth's surface in Crete, which thereby generated a huge tsunami which caused extensive damage around the eastern Mediterranean. Similarly, Krakatoa erupted in 1883. These volcanic incidents left many dead directly due to the fallout, but also because of the consequent affect on farming around the world for a few years afterwards.

Chapter 7: The Flood

[57] As mentioned above, it is very difficult to estimate the potential rise in sea level if all the ice and snow on the Earth melts

[58] The Epic of Gilgamash was written in about 612BC. It tells the story of a massive flood that devastated a huge area, and could well have formed a text that was used by the writer of the Noah account. See the Wikipedia article on The Epic of Gilgamash (accessed on 05-11-2013).

[59] See the Wikipedia article on the History of Babylon by Berossus (accessed on 05-11-2013).

[60] See Plimer ibid p296

[61] See particularly Chapters 11, 25, 32, 33 and 34

[62] Peter Harris' contribution in 'Caring for Creation', ed S Tillett, Bible Reading Fellowship, p 49

Chapter 8: Water in arid areas

[63] The MDGs include the supplying fresh water to more than one billion people, and providing effective sanitation to more than 2 billion people

[64] For a more comprehensive introduction to the Healthy Vine, see The Healthy Vine Trust, http://www.healthy-vine.org/water.html

[65] Nuclear fusion is viewed as a solution to humankind's energy problems if the plasma seeded with tritium can be maintained at a temperature above 10^6 °C, which implies it has to be contained in a vacuum without the plasma touching the side of the containing vessel. See the Wikipedia articles on Nuclear Fusion and Joint European Torus (accessed on 05-11-2013)

[66] See the Wikipedia article on Desalination (accessed on 05-11-2013).

Chapter 9: Water and the Exodus

[67] Scholars have a number of different views of where the Israelites crossed the 'Red Sea', but one of the more popular routes is to the north of the delta where the Suez Canal has been built.

Chapter 10: Water from the Rock

[68] For the possible origins of sources 'J' and 'P' for the text of Genesis, see the Wikipedia article on the 'Documentary Hypothesis' (accessed on 05-11-2013).

[69] Benares is the chief centre of Hindu pilgrimage

Chapter 11: Water and ritual washing

[70] For the translation of the Quran by Yusuf Ali see https://www.quranonline.net/html/trans/options/yali/21.html (accessed on 05-11-2013).

[71] For more on Lourdes see the associated Wikipedia article (accessed on 05-11-2013).

[72] On the Day of Atonement (Leviticus 16) a live goat (the scapegoat) was to carry the sins of all the people outside the camp of the community into the desert where it was released and set free.

Chapter 12: River Jordan and the Dead Sea

[73] Prof Patrick Denny was Professor of Wetland and Aquatic Ecology at UNESCO-IHE, 1994-2004

[74] See the Wikipedia article on the Dead Sea (accessed 01-11-2013)

[75] See the Wikipedia article on the Sea of Galilee (accessed 01-11-2013)

[76] See the Wikipedia article on the thermal springs at Hammat Grader (accessed on 01-11-2013

[77] See the Wikipedia article on the Dead Sea scrolls (last viewed on 01-11-2013). Also, refer to some of the many books on the scrolls. One of the first was written by Edmund Wilson (1955). The Scrolls from the Dead Sea,

W H Allen, which is a lecture he gave following the discovery of the scrolls by a Beduin boy in the Spring of 1947.

[78] See the Wikipedia Article on the Allenby Bridge (accessed on 05-11-2013).

[79] We do not look in detail at the consequences of the IT revolution on water management, and the Christian appreciation of water. A number of philosophers have considered the implications of technology for human beings and societies. For an introduction to the subject, see Abbott, M B (1991) Hydroinformatics: Information Technology and the Aquatic Environment. Aldershot, UK / Brookfield USA: Ashgate, and primers on Heidegger and Jaques Ellul

Chapter 13: Crossing the Jordan

[80] I have The Netherlands in mind!

[81] For information on sales of bottled water and associated issues see the Wikipedia article on 'Bottled Water' (accessed on 05-11-2013).

[82] See the Wikipedia article on bottled water (accessed on 01-11-2013)

[83] See the web site of the Natural Resources Defense Council, a USA Environmental action group

[84] See the Wikipedia article on plastic bottles (accessed on 01-11-2013)

Chapter 14: A cup of water

Chapter 15: Water and hospitality

Chapter 16: Water and urban society

[85] The United Nations has stated that 50% of the world's population was in cities by 2008. In its World Water Development Report Number 4 UNESCO states that 'by 2030 about 60% of the world's population is expected to live in urban areas' p31

[86] See the discovery of the Siloam Inscription in Easton's Bible Dictionary at the Christian Classics Ethereal Library

[87] For further information on water supply to Jerusalem in the Maccabean era see the Wikipedia article on Jerusalem during the Second Temple Period (accessed on 05-11-2013). (accessed on 05-11-2013).

[88] For the present day water supply system operated by Mekorot in Jerusalem see http://www.mekorot.co.il/Eng/Mekorot/Pages/IsraelsWaterSupplySystem.aspx (accessed in 05-11-2013).

[89] Water carriers are still very prominent in some Middle East and Mediterranean countries, including Morocco, though bottled water now limits their role to being a tourist attraction.

[90] Ibid p 25

[91] Hence the term 'Public Health engineering'

[92] Butler, D and Davies, J W (2011) Urban drainage, 3rd Edn, Spon Press

[93] Jean-Pierre Coubert: 'The Conquest of Water' Princeton University Press

[94] Price, R K and Vojinovic, Z (2011) Urban Hydroinformatics. IWA Press

[95] Vojinovic Z and Abbott M B (2012) Flood risk and social justice. IWA Press

[96] The Millennium Development Goals (MDGs) were established by the United Nations in 2000 with the commitment to achieve them by 2015. They are stated in the United Nations Millennium Declaration (2000)

Chapter 17: Bitter waters made sweet

[97] Martin Palmer (2010) The river Jordan's Shame. Guardian Newspaper, Tuesday 27 July 2010

[98] See the Wikipedia article on the Allenby Bridge (accessed on 01-11-2013)

[99] Qasar al-Yahud is a possible site of Jesus' baptism

[100] Wadi Kharrar, or the Pools of Elijah, are also a possible site where Jesus was baptised

[101] Alliance of Religions and Conservation

[102] Urban Wastewater Treatment Directive, Council Directive 91/271/EEC, 1991, and Water Framework Directive, Council Directive 2000/60/EC, 2000

[103] Butler, D and Davies, J W (2011) Urban drainage, 3rd Edn Spon Press

[104] See Ian Bradley, 'A Spiritual of History of Water' Bloomsbury

[105] Bradley ibid Chapter 1

[106] Catholic initiatives in North America; see http://www.earthlight.org/essay41_columbia_river.html

[107] For information on the results of the European Union research project SWITCH, see http://ec.europa.eu/research/headlines/news/article_13_05_27_en.html (accessed on 05-11-2013).

[108] For natural treatment of wastewater using constructed wetlands see the associated Wikipedia article (accessed on 05-11-2013).

Chapter 18: As the waters cover the sea

[109] TearFund, Churchgoing in the UK, published in April 2007

[110] See the Wikipedia article on Higg's Boson. This is an elementary particle which is fundamental to the Standard model and other theories in

Particle physics. The particle has been dubbed the 'God particle' (accessed 01-11-2013).

[111] See the Wikipedia article on the 'Eastern Scheldt Storm Surge Barrier' (accessed 01-11-2013). The barrier was commissioned in 1986.

Chapter 19: All that live in the waters

Chapter 20: Water and character formation

Chapter 21: Dew and grace

Chapter 22: Water and God's Word

Chapter 23: By the waters of Babylon

[112] See for example Rivers of Babylon by Sublime

[113] See the Wikipedia article on qanats (accessed on 5-11-2013)

[114] We know much about the Gardens through accounts by Berrosus, who is quoted by such writers as Flavius Josephus writing in the 2nd and 1st century BC and early 4th century BC texts of Ctesias of Cnidus (see the Wikipedia article on the 'Hanging Gardens of Babylon')

[115] Establishment of modern Israel was on May14, 1948 with support from United States, United Kingdom and Russia.

[116] Wright p 201

[117] Ibid p 176

Chapter 24: Water and birth

[118] For information on how the kidneys function in the human body see the corresponding Wikipedia article (accessed on 05-11-2013).

[119] For average reservoir residence times of water see the Wikipedia article on the Water Cycle (accessed on 05-11-2013).

Chapter 25: Water and blood

[120] See the Wikipedia article on Blood (accessed on 05-11-2013).

[121] See the Invitation to Communion, Common Worship: Services and Prayers for the Church of England. Church House Publishing, p 180

Chapter 26: Water into wine

[122] Drower, E S (1956) Water into Wine: A study of ritual idiom in the Middle East. John Murray, London

Chapter 27: Living water

[123] For more information on the Healthy Vine Trust at Sekamuli, see http://www.healthy-vine.org/ (accessed on 05-11-2013).

[124] For information on the trick water fountains at Schloss Hellbrunn in Salzburg see http://www.hellbrunn.at/en/ (accessed on 05-11-2013).

Chapter 28: Into deep water

Chapter 29: Walking on water

[125] For information on Leonardo DaVinci's fascination with water see Professor Chris Whitcombe's web site on http://witcombe.sbc.edu/water/artleonardo.html (accessed on 05-11-2013).

[126] Young, W P (2007) 'The Shack'. Windblown Media

Chapter 30: Calming the storm

[127] For further information on the Superstorm Sandy (or formally hurricane Sandy) see the Wikipedia article on 'Hurricane Sandy' (accessed on 05-11-2013).

[128] There are a number of articles and reports on the 1953 North Sea disaster: for example, the Report on the Conference organised by the Institution of Civil Engineers in December 1953, and the Wikipedia article 'North Sea Flood of 1953' (accessed on 05-11-2013).

Chapter 31: Water and healing

[129] See the Wikipedia article on 'Regeneration' (accessed on 05-11-2013)

[130] See the Wikipedia article on 'Hansen's disease' (accessed on 05-11-2013).

[131] For information from the World Health Organisation on diarrhoeal disease see the site http://www.who.int/water_sanitation_health/diseases/diarrhoea/en/ (accessed on 05-11-2013).

[132] See the Wikipedia article on 'Spa, (accessed on 05-11-2013).

[133] See the Wikipedia article on 'Roman Baths (Bath)' (accessed on 05-11-2013).

[134] For further information on Bad Gradierhaus see the site http://www.
bavaria.by/bavarian-gardens-parks/a-royal-spa-garden-bad-reichenhall
(accessed on 05-11-2013).

[135] Spa water discovered; see John Winsor Harcup: the Malvern Water Cure,
p20

[136] Bradley ibid, see Chapter 3 and following chapters

[137] Homily on baptism delivered on the Feast of Epiphany in 381 AD by St
Gregory of Nazianus

[138] See 'Early Christian Writings' translated by Maxwell Staniforth, Penguin
Classics p81

Chapter 32: Water and the Cross

[139] Lady Drower on the Eucharist: ibid p 91 and following

Chapter 33: Waters of Baptism

[140] Mandaean baptism is related to the baptism of John the Baptist

[141] Lady Drower ibid p 15 *and following*

Chapter 34: The gift of tears

[142] Hausherr, I SJ (1891) Penthos: The Doctrine of Compunction in the
Christian East. Translated by Anselm Hufstader. Cistercian Publications
Kalamazoo, Michigan

[143] For a view on the God-is-dead movement see the corresponding
Wikipedia article on the subject. (accessed on 05-11-2013).

[144] Keller T (2009) 'The Reason for God'. Riverhead Books

[145] Huasherr, ibid, p23

Chapter 35: Down to the sea

[146] Hendrik Willem Mesdag (23 February 1831-10 July 1915)

[147] Scheveningen harbour was constructed in 1904

[148] Sir Francis Charles Chichester (17 September 1901-26 August 1972), became
the first person to sail single-handed around the world by the clipper
route, and the fastest circumnavigator, in nine months and one day
overall. See the corresponding Wikipedia article (accessed 05-11-2013).

[149] Sir Edmund Percival Hillary (20 July 1919-11 January 2008), a New Zealand
mountaineer, explorer and Nepalese Sherpa mountaineer Tenzing Norgay
became the first climbers confirmed as having reached the summit of

Mount Everest on 29 May 1953. See the corresponding Wikipedia article (accessed on 05-11-2013).

[150] The Titanic was a British passenger liner that sank in the North Atlantic Ocean on 15 April 1912 after colliding with an iceberg during her maiden voyage from Southampton, UK to New York City, US. See the corresponding Wikipedia article (accessed on 05-11-2013).

[151] The 2004 Indian Ocean tsunami was generated by an undersea earthquake that occurred on Sunday, 26 December 2004, off the west coast of Sumatra, Indonesia. See the corresponding Wikipedia article (accessed on 05-11-2013).

[152] The 2011, Great East Japan Earthquake, on 11 March 2011 was the most powerful earthquake to hit Japan, and generated 40.5m high tsunami waves that devastated Miyako in Tōhoku's Iwate Prefecture. The waves travelled up to 10km inland in the Sendai area. See the corresponding Wikipedia articles (accessed 05-11-2013).

[153] The AD 365 Crete earthquake occurred on 21 July 365 in the Eastern Mediterranean, with epicentre near Crete and followed by a tsunami which devastated the southern and eastern coasts of the Mediterranean, particularly Libya, Alexandria and the Nile Delta. See the corresponding Wikipedia article (accessed on 05-11-2013).

Chapter 36: Water in the Middle East

[154] Most people regard the Nile as the longest river on the planet, but it all depends on the identification of the source and the mouth of a river. The Amazon has by far the largest flow at its mouth

[155] Mekorot, founded in 1937, is the national water company of Israel

[156] For more information on the West Bank aquifers see http://mideastweb. org/westbankwater.htm (Accessed on 05-11-2013)

[157] World Bank report 'Assessment of Restrictions on Palestinian Water Sector Development' (2009) available at http://siteresources.worldbank. org/INTWESTBANKGAZA/Resources/WaterRestrictionsReport18Apr2009. pdf (accessed 05-11-2013).

[158] For more information about the Israeli - Palestinian conflict see the Wikipedia article at http://en.wikipedia.org/wiki/ Israeli%E2%80%93Palestinian_conflict

[159] See the Wikipedia article on the King Abdullah Canal east of the Jordan river in Jordan (accessed on 05-11-2013)

[160] United Nations Convention on the Law of the Non-Navigation Uses of International Watercourses can be accessed at http://legal.un.org/avl/ha/clnuiw/clnuiw.html

[161] 1998 water act in South Africa can be accessed at http://www.dwa.gov.za/Documents/Legislature/nw_act/NWA.pdf

[162] Canadians for Justice and Peace in the Middle East (CJPME) can be accessed at http://www.cjpme.org/

[163] See, for example, The International Covenant on Economic, Social and Cultural Rights (1966) Articles 1, 11, 25; The Convention on the Elimination of All Forms of Discrimination against Women (1979) Article 14-2(h); and The Convention on the Rights of the Child (1989) Article 24-2(c).

[164] CJPME Ibid

[165] Palestine observer status: see the Wikipedia article on 'United Nations General Assembly observers'

Chapter 37: Virtual water

[166] See Virtual Water data at http://www.trust.org/item/?map=factbox-how-much-virtual-water-do-you-use-every-day/ (accessed on 05-11-2013).

[167] See the Water Footprint organization site at http://www.waterfootprint.org/?page=files/home (accessed on 05-11-2013).

[168] Hoekstra A Y and Hung P Q (2002) Virtual Water Trade. UNESCO-IHE, Delft

Chapter 38: Water and natural disasters

[169] While doing the final editing of this book the typhoon Haiyan hit the Philippines. With winds in excess of 210 miles / hour the devastation was enormous. Thousands were killed and hundreds of thousands were made homeless. As time went on it became more apparent that disease and starvation were threatening the population directly affected. In particular, there was very little safe drinking water. The survival of thousands was highly dependent on the response of the world's disaster relief agencies. Without a concerted and coordinated programme of aid many more would die after the typhoon had long gone than during its brief impact on the scattered islands of the Philippines, devastating as it was.

[170] The Millennium is a very important topic for some Christians. As it does not appear to depend significantly on water, this topic is not reviewed in this book.

[171] The work of Walton suggests that with no darkness (night) following day (light) there is no process of time

Water Acts

[172] Geddes, P (1915). *Cities in Evolution*. London: Williams. The phrase 'Think globally, act locally' does not appear explicitly in this book, but it is clearly embedded in the text.

Glossary

Aquifer: an underground layer of water-bearing permeable rock from which groundwater can be extracted using a well.

Brownian movement: the random movement of particles suspended in a fluid resulting from their bombardment by fast-moving atoms or molecules in the fluid.

Canaan: a historical Semitic-speaking region roughly corresponding to modern-day Israel, Palestine, Lebanon, and parts of Jordan and Syria

Carcinogenic: causing or tending to cause cancer

Climate change: a significant and lasting change in the statistical distribution of weather patterns over periods ranging from a decade to millions of years.

Cloud, The: the use of computing resources that are delivered as a service over the Internet.

Computer simulation modeling: a computer program that attempts to simulate an abstract model of a particular system.

Coriolis force: a deflection of moving objects (or water in the oceans) when viewed in a rotating reference frame.

Cosmos: is the universe regarded as an ordered system.

Creation: creation ex nihilo, the concept that matter comes 'from nothing'.

Density: mass per unit volume.

Desalinization: the removal of salt or other chemicals from seawater, achieved by means of evaporation, freezing, reverse osmosis, ion exchange, and electrodialysis.

Effluents: an outflow of water or gas from a natural body of water, or from a man-made structure.

El Nino: a band of anomalously warm ocean water that occasionally develops off the western coast of South America and is associated with climate change.

Fall, The: the breaking of the relationship between Adam and Eve and God through Adam and Eve's disobedience, with far-reaching consequences for the human race and the physical universe.

Flood, The: A catastrophic flood over the whole earth that destroyed all land-based animal life except for Noah and his family and male and female of each animal species who were saved by residing in an ark.

Global warming: the perceived rise in the average temperature of the Earth's atmosphere and the oceans since the 20th century and its projected continuation.

Greenhouse gas: a gas in an atmosphere that absorbs and emits radiation within the thermal infrared range.

Gulf stream: a powerful, warm, and swift Atlantic ocean current that originates at the tip of Florida, and follows the eastern coastlines of the United States and Newfoundland before crossing the Atlantic Ocean towards Europe.

Heavy metal: a member of a loosely defined subset of elements that exhibit metallic properties.

Hebrews: the Israelites, especially in the period before they had settled in Canaan and when they were still nomadic.

Hemoglobine: a protein that occurs in the blood of the human and many other animals.

Higg's boson: a basic elementary particle initially postulated in 1964, and tentatively confirmed to exist on 14 March 2013.

Humidity: the amount of water vapor in the air.

Hydrological cycle: describes the continuous movement of water on, above and below the surface of the Earth.

Hydroinformatics: a branch of informatics which concentrates on the application of information and communications technologies (ICTs) in addressing the equitable and efficient use of water for many different purposes.

Hydromorphological resistance: the tendency of a natural catchment to retain water from precipitation which is reduced if trees are removed or streams are cleared and straightened

Incompressibility: a flow in which the material density is constant within a fluid parcel

LIDAR: **LI**ght **D**etection **A**nd Ranging (**LIDAR**), (sometimes **Laser Imaging Detection And Ranging**) is an optical remote sensing technology that can measure the distance to, or other properties of, targets by illuminating the target with laser light and analyzing the backscatter.

Littoral drift: consists of the transport of sediment along a coastline by a current generated by waves at an angle to the shore, dependent on the prevailing wind direction, swash and backwash.

Long wave: a wave on the interface between two fluids such that the length of the wave is considerably greater than the depth of the denser fluid.

Melting and boiling points: the temperature at which a substance changes state from solid to liquid.

Microtubules: are components of the cytoskeleton, found throughout the cytoplasm of a biological entity.

Millennium Development Goals (MDGs): are eight international development goals that were officially established following the Millennium Summit of the United Nations in 2000, following the adoption of the United Nations Millennium Declaration.

Millibar: a unit of atmospheric pressure equal to one thousandth (10^{-3}) of a bar. Standard atmospheric pressure at sea level is about 1,013 millibars.

Nano technology: the manipulation of matter on an atomic and molecular scale.

Non-Government Organisation (NGO): is a legally constituted organization created by natural or legal people that operate independently from any form of government.

Potable water: water safe enough to be consumed by humans or used with a low risk of immediate or long term harm.

Primary, secondary and tertiary treatment: the process of removing contaminants from wastewater and household/sewage.

Qanat: a water management system used to provide a reliable supply of water for human settlements and irrigation in hot, arid and semi-arid climates.

Receiving waters: a river, ocean, stream, or other watercourse into which wastewater or treated effluent is discharged.

Rip current: a strong plume of water flowing seaward from near the shore, typically through the surf line.

Redemption: The restoration of man from the bondage of sin to the liberty of the children of God through the satisfactions and merits of Jesus Christ.

Runoff: the flow of water, from rain, snow melt, or other sources, over land.

Sewage: a water-carried waste, in solution or suspension that is intended to be removed from a community.

Short waves: a wave at the interface between two fluids (or the same fluid at different densities) where the length scale of the wave is considerably shorter than the depth of either fluid.

Solvent: a substance that dissolves a solute (a chemically different liquid, solid or gas), resulting in a solution.

Specific heat capacity: the measurable physical quantity that gives the amount of heat required to change the temperature of a substance by a prescribed amount.

Spring: is any natural situation where water flows to the surface of the Earth from underground.

Stade: ancient unit of distance measurement.

Stormwater: water that runs off over the ground surface during a precipitation event and may collect and cause damage unless conveyed away intentionally in a network of channels and/or pipes underground.

Surface tension: a contractive tendency of the surface of a liquid that allows it to resist an external force.

Storm surge: an offshore rise of water associated with a low pressure weather system, typically a tropical cyclone or a strong extratropical cyclone.

Superstorm: a subjective term for any storm that is extremely and unusually destructive.

Tabernacle: the portable dwelling place for Yahweh from the time of the Exodus from Egypt through the conquering of the land of Canaan.

Tsunami: a series of water waves caused by the displacement of a large volume of a body of water, typically an ocean or a large lake by an earthquake, volcanic eruption, landslide, glacier calving, meteorite impact or other disturbance.

Turbulence: a flow regime characterized by chaotic and stochastic property changes.

Undertow: a subsurface flow of water returning seaward from shore as result of wave action.

Viscosity: a measure of a fluid's resistance to gradual deformation by shear stress or tensile stress.

Washoff: the erosion or pickup of loose material on the catchment surface by runoff.

Wastewater: water that contains waste, in solution or suspension, and is intended to be removed from a community.

Verses quoted from the Bible involving water

Genesis
1
1:2
1:9
1:9-10
1:11
1:23
1:26
1:28
2
2:4-25
2:7
2:10
2:14
2:24-25
3:1-24
3:16
6:1-8:22
6:7
7:11
8:22
9:1
9:4-5
9:8-17
9:16
15:18
16:6
18
18:1-15
19:21-22
24
26:14-17
26:15
26:17

Exodus
15:22-26
15:26
17:1-7
17:7
24:8
25:1-18
29:4
30:17
Leviticus
1:9
6:27
8:21
14:5
14
14:5
14:35-57
15
17:11
Numbers
4:6
19
20:1-13
Deuteronomy
1:7
11:10
29:3
29:5
29:23
Joshua
1:4
3
3:15
4

Judges
6:33-7:25
7:17-22
15:19
Ruth
1 Samuel
2 Samuel
23:8-17
1 Kings
9:6-28
10:22
19:6
2 Kings
2:19-22
18:17
20:20
1 Chronicles
2 Chronicles
32:3-5,30
32:4,10
32:27-33
Ezra
Nehemiah
Esther
Job
5:10
6:15,16
10:8
10:17
12:15
14:11
14:12
14:5-31
22:7

Jude
Revelation
1:5-6
5:4
7:17
8:6-11
10
13:17
13:1
14:6
16
16:12
19:11
20:13
21:1
21:4
21:6
22
22:1-2
22:2
22:5
9:5
12:11
22:14

Book and website list

The references identified here are only a few of the many relevant books and websites that could have been included.

United Nations Development Program—World Water Development Report
IPCC (2013) 5th Assessment Report'. See http://www.ipcc-wg2.gov/SREX/
Managing Water under Uncertainty and Risk, 2012.
Water in a Changing World, 2009.
IPCC (2007) '4th Assessment Report''. See http://www.ipcc.ch/
Previous editions, 2006 & 2003. See http://www.ipcc.ch/

Virtual water
Water Footprint Network. (2011) Water Footprint Assessment Manual, EarthScan
Global Water Footprint Standard.
http://www.waterfootprint.org/?page=files/GlobalWaterFootprint
Water Footprint Network: WaterStat.
http://www.waterfootprint.org/?page=cal/WaterFootprintCalculator

NGOs
AquaForAll
A-Rocha: http://www.arocha.org/gb-en/index.html
Tearfund, http://www.tearfund.org/About+us/What+we+do/Improving+basic+services.htm
The Healthy Vine Trust, http://www.healthy-vine.org/water.html
WaterAid: http://www.wateraid.org/uk/

Relevant websites
Barack Obama speeches: http://obamaspeeches.com
Lao Tzu quotes: http://thinkexist.com/quotes/lao_tzu/. See also), en.wikipedia.org/wiki/Laozi
'Guide me Oh thou great redeemer' by William Williams, http://en.wikipedia.org/wiki/Cwm_Rhondda
John L Cullney quote from Wilderness Conservation 1990, quoted in http://www.cybergeo.com/wellness/water/quotes.html

'The Church's One Foundation' by Samuel John Stone, in 'Lyra Fidelium;
Twelve Hymns on the Twelve Articles of the Apostles' Creed.'.\; see http://en.wikipedia.org/wiki/The_Church's_One_Foundation
Leominster Parish Church www.leominsterpriory.org.uk

Books written from an overall Christian perspective
Bailey, K.. (2008) 'Jesus through Middle Eastern eyes: Cultural studies in the gospels'. IVP Academic
Berry, R. J. (2005) 'Rejection of the Creator', in 'Caring for Creation', edited by S Tillett, Bible Reading Fellowship
Biswas, A. K. (1970) 'History of Hydrology', North Holland Publishing Company, Amsterdam
Bradley, I. (2012) 'Water: A spiritual history'. Bloomsbury.
Burke III, E. and Pomeranz, K. (2009) The environment and world history'. University of California Press
Didache, early Christian teaching document
Donne, J. 'Selected Poems' edited by John Carey (1996) Oxford University Press
Drane, J. W. (1990) 'An Introduction to the Bible', Lion Publishing
Drower, E. S. (1956) 'Water into Wine: A study of ritual idiom in the Middle East'. John Murray, London.
Dyke, F. van et al, (1996) 'Redeeming Creation: The biblical basis for environmental stewardship'. IVP
Finkel, Irving (1988) 'The Hanging Gardens of Babylon,' in 'The Seven Wonders of the Ancient World', edited by Peter Clayton and Martin Price, Routledge, New York, pp. 38 ff
Hallman, D. G, (2000) 'Spiritual values for earth community'. WCC Publications Geneva
Hausherr, I, (1891/1982) 'Penthos: The doctrine of compunction in the Christian East'. Cistercian Publications.
Lewis, C. S. (1947) 'Miracles'. Collins, Fontana Books
McGrath, A. P. (2010) 'A Life – a biography of C. S. Lewis'.
Mello, A. de (1989) 'The Heart of the Enlightened'. Fount, Harper Collins
Ortberg, J. (2001) 'If you want to walk on water - you have to get out of the boat!!' Zondervan

Petersen, E. (2005) 'Creation and the gift of time', in 'Caring for Creation', edited by S Tillett, Bible Reading Fellowship

Polkinghorne, J. (1994) 'Science and Christian Belief', SPCK

Richards, J (2001) 'Tears - Gift of the Spirit?' Available at www.helpforchristians.co.uk

Rossetti, Christina Georgina, 'By the Sea' In (2008) 'Poems and Prose', Oxford University Press

Ross, M. (1987) 'The Fountain & the Furnace: The way of tears and fire'. Paulist Press

Schaeffer, F. A. (1970) 'Pollution and the death of man: The Christian view of ecology'. Hodder and Stoughton

Tillet, S. (editor), (2005) 'Caring for Creation'. Bible Reading Fellowship

Walton, J. H. (2009) 'The lost world of Genesis One'. IVP Academic

Ware, Bishop Kallistos (1979) 'The Orthodox Way'. A. R. Mowbray & Co. Ltd.

Wright, N T (2012) 'Simply Jesus'. SPCK

Young, W. P. (2007) 'The Shack'. Windblown Media

Other books

Updike, J. (1989) 'Self-Consciousness: Memoirs', Random House Trade Paperbacks

Brown, L. A., Renner, M. and Halweil, B. (1999) 'Vital Signs 1999'. W. W. Norton, New York

Eliot, T. S. (1944) 'The Dry Salvages' in 'Four Quartets'. Faber & Faber, London

Kron, W. (2013) 'Flooding – There is no such thing as complete protection'
from 'Topics Online', MunichRE, 05-07-2013

Rechberger, K. (2009) in 'Water-management – A Global Innovation Outlook Report', IBM

Viliers, M. de (2001) 'Water: The Fate of Our Most Precious Resource'. First Mariner Books, for quotation from Levi Eshkol, Israeli Prime Minister, 1962

See http://www.waterfootprint.org/?page=cal/WaterFootprintCalculator

Climate change

Plimer, I. (2009) 'Heaven and Earth: Global warming: the missing science'. Taylor Trade Publishing

IPCC (2013) 5th Assessment Report. http://www.ipcc.ch/
Garvey, J. (2008) 'The ethics of climate change: Right and wrong in a warming world'. Continuum

Hydrology
Biswas, A. K. (1970) History of Hydrology, North Holland Publishing Company, Amsterdam
Coubert, J-P. The conquest of water: The advent of health in the industrial age. Princeton (translated by Andrew Wilson)
Burke III, E. and Pomeranz, K. (2009) The environment and world history. University of California Press

About the Author

It is only right to give you, as reader, some indication of the reasons for my enthusiasm for water in the environment and its management. To do this, I describe here some particular experiences from my childhood, and subsequently from my student years and professional life. Like many children, my first enjoyable memories of water were at the sea-side, attempting to build sandcastles that would defy the persistent encroaching waves, or of being turned upside down by breakers on the sandy beach. I welcomed the challenge of trying to dam streams in the Welsh hills or to skim flat stones across the river at Hay-on-Wye, where my paternal grandparents lived. But water at the seaside was not just an enjoyable experience. I became aware of the awesome power of water when news came of serious flooding from a disastrous surge in the North Sea along the East Anglian coast in 1953, and the unacceptable and devastating floods from the River Lugg invaded the lower part of Leominster, the small country town where I was brought up. Little did I realise then that these events would be seminal in forming the focus of my future career.

In my last year at the local Grammar School I bought a book on physics that discussed the flow of fluids. This made me aware that there was some interesting mathematics undergirding the subject, but it seemed complicated and abstruse, depending on partial differential equations, which were then a considerable mystery to me. I was accepted to study maths at Cambridge, but it was not until my third year that I was able to begin to unlock the mysteries of fluid dynamics. My fourth year doing Part III of the Maths Tripos (the Cambridge equivalent of an MSc) was given over to a range of topics in fluids: aeronautics, magneto-hydrodynamics, meteorology, turbulence, open channel flow, oceanography, cosmology etc. After some uncertainty, I resolved to do a PhD in fluid dynamics. One of the lecturers in the subject at Cambridge, Dr Ian Proudman, was appointed Professor of Mathematics at the completely new University of Essex, so I jumped at the chance of joining him as one of his graduate students in the autumn of 1964.

For the first year I explored the delights of Ekman layers in deep oceans and other strange phenomena, but these had been researched by others and I could not find a research problem that was potentially tractable. Then Prof. Proudman, suggested the topic of breaking waves. The die was cast, and I enthusiastically set to work. For the next four years I looked at the problem from a range of different (mathematical) points of view. I spent hours on holiday at beaches fascinated and entranced by the waves curling over at the top and then 'breaking'. I would like to think that I made a significant contribution to the subject, but the problem is highly non-linear, and mathematical techniques for this sort of problem are very limited. Although I had done enough for a PhD, I needed a new approach to the problem. Solving the non-linear mathematical equations for fluid flow analytically, even under very simplified conditions, was virtually impossible. But there was another way of tackling the problem— using computers. These had become commercially available during the early 1960s, and the new University of Essex had an ICL 1900 series machine. My fellow research students and I began to see the possibilities of the electronic computer in helping us to solve our non-linear equations numerically. So in 1967, with the aid of a Fellowship, I began to solve my breaking wave equations on the computer. I managed to get to the stage where the wave was beginning to curl over: I had begun to conquer the process.

But I was restless. Detailed numerical calculations of one breaking wave were interesting from a theoretical point of view, indeed, the calculations were fascinating. But what were the practical implications of this solution? I could not break my virtual wave against a virtual wall and look at the consequent pressures as would interest an engineer, or look at the effect on a breaking wave of the returning surge back down the beach from the previous wave. I was focussed so much on the detailed dynamics that what I was doing was of little practical use. I needed a change of attitude. Five years of analytical mathematics was enough. I decided to get involved in computer based hydraulic engineering.

The opportunity came to do research for a year at the Coastal and Oceanographic Engineering Department at the University of Florida in Gainesville. There, accompanied by my wife Thea, I became involved in exploring hurricane surges (analytically!) and modelling heated outfall discharges from proposed nuclear power stations in shallow coastal waters (numerically!). Although I enjoyed the challenge of this research, I felt called to be more involved in Christian ministry.

So far I have said little about my awareness of the Christian faith and its place in my early life. My first memory of church was being taken to Brecon cathedral as a child on Easter day and being entranced with the Easter garden made entirely out of spring flowers. Then at the age of seven I became a chorister in the large Parish church of St Peter and St Paul in Leominster. I was fortunate to have a wonderful choir master who taught me the rudiments of music, and when I was older, the organ. I remained a member of the choir until I went to university at Cambridge to read mathematics. During my mid-teenager years, whereas I enjoyed my singing in the church choir, I felt that I was missing out on something deeper in being a Christian than I had experienced. This was particularly highlighted by the strong personal faith centred in Jesus Christ which was exhibited by my non-conformist friends. Their vibrant faith persuaded me that when I went to Cambridge I would find out more about Jesus. So it was in October 1960 I was invited by some members of the Christian Union to the 'Freshers' sermon' which was being given by David Shepherd, the former England batsman now a vicar in the Church of England[1]. His preaching challenged me to commit my life to Christ, and I did that night. I have spent my life since then working out what this means for me, and the journey has been one that has largely focussed on being part of and helping to develop strong local church communities with underlying charismatic gifts in a village south of Oxford in England, and in The Hague in the Netherlands.

The stirrings towards Christian ministry that I experienced in Gainesville, Florida, led Thea and I to return to England where I spent some time exploring ordination to priesthood in the Church of England. This did not happen then, and instead I joined a UK government laboratory: the Hydraulics Research Station in Wallingford. I now became a mathematician in a truly engineering environment; at long last I could hopefully put my skills to better practical use.

---

[1]   David Shepherd later became Bishop of Liverpool, and wrote a moving autobiography: 'Steps along Hope Street', Darton, Longman and Todd

In view of my experiences in Florida, I expected to work on coastal problems; but I was asked to investigate methods for flood propagation in rivers as part of a UK Flood Studies project. This enabled me to do some analytical mathematics as well as computational hydraulics. Later, I became involved in the computer simulation of flows in urban drainage systems consisting of networks of pipes under the ground. The objective was to assess the performance of networks that had been designed on the computer for alleviating flooding. As so often happens, success in one area leads to new opportunities of addressing problems in another. In particular, I became involved in applying the computer simulations to determine how best to rehabilitate drainage networks that were deteriorating with age and usage. The tools I was developing were also being made available (at a price) to river and urban drainage engineers in the UK.

From 1982, following the privatization of the laboratory where I worked, my colleagues and I were free to develop the commercial potential of the modelling software products. So I spent the next 12 years building up a subsidiary business to market and further develop such products. But although I enjoyed the challenge of building up a successful commercial business from scratch, I wanted to return to research: others could do a better job of managing the commercial aspects of the business than I could. The opportunity came following a significant disappointment. I had become enamored by the use of expert systems and artificial intelligence to analyse scientific texts, and I wanted to use related techniques to improve the knowledge provided on-line to engineers using modelling software products. Khurshid Ahmad, a colleague at Surrey University and I, set up a second business venture in HR Wallingford that in the first instance provided translation tools to translators of technical texts. It turned out, however, that this was too far outside the core business of my parent company, and the main board decided not to go any further with the business. The day after receiving the news that the second business venture was to be closed down, I was invited to apply for the post of Professor of Hydroinformatics at IHE in Delft.

Crossing boundaries, whether physical or mental, is something that most of us have to face at one stage or another in our lives. I was happy working at HR Wallingford in the commercial software house that I had helped to establish. In deciding to put my name forward for the post of Professor of Hydroinformatics at IHE Delft (presently an important UNESCO Category 1 post-graduate institute) I knew I was crossing a real physical sea boundary that separated England from the Netherlands, and that Thea and I would have to face up to living in another country. We had been married for almost thirty years and our youngest children were well into their teens. Up to then we had no expectation of moving from the village where we lived. But God had other ideas, and the post in Delft was offered to me. There were not a few tears shed in making the decision to move, not least because we were leaving behind many dear friends who had been a key part of our life in the village. Although we knew that many people go and live in countries other than the one they were born in, this did not minimize the fact that for us it was still a major adventure. Being aware that God was with us in the move helped us make the transition. The last fifteen years have been ones that I would not have missed, especially as they have brought me in touch with many wonderful people from all over the world. I have learned much about the practical aspects of water and its management, and have had the privilege of working with students from many different countries doing research in hydroinformatics (the management of water in the environment using information and communication technology).

This academic post gave me opportunities for research in water management that I had only dreamt of, particularly in developing countries. Coming from industry rather than academia, it took time for me to build up a research group and accrue a sufficient number of peer-reviewed publications, but by the time of my retirement I had promoted and worked with 14 PhD students doing research into: uncertainty in model predictions, data driven modelling, chaos, anticipatory water management, information theory for monitoring, urban flood modelling, inclusion of real time control in urban drainage design, providing water services for urbanising areas, digital terrain models from remote sensing, and eutrophication in deep reservoirs and shallow lakes, supported by colleagues from all over the world. Doing this research, and establishing a regular Masters course in hydroinformatics for 15-20 mature students annually from developing countries, was my privilege during the 10 years I was in full-time employment at UNESCO-IHE (we became a UNESCO Institute in 2007). I could not have asked for a better way to conclude my professional career. Even in retirement, I have continued developing software for flooding in rivers and urban areas and writing text books for researchers.

So now you know something of my professional background. On the spiritual front, I was a Lay Reader in the Church of England from 1968 to 1995, at which time I was ordained deacon in Christchurch, Oxford and priested in 1996. All the while I have been intrigued by the references to water in the Bible and how faith communities can contribute to the achievement of the Millennium Development Goals. My aim in this book is to take you on a journey so that I can share with you some of the challenges and opportunities there are for deepening our understanding of water in creation and of God's plan for our redemption, and to encourage you to develop a personal relationship with Jesus Christ. I invite you to put on your spiritual life-jackets!

Roland Price
The Hague, November 2013

Index

W

Y

Z

Lightning Source UK Ltd.
Milton Keynes UK
UKOW03f1801090614

233119UK00001B/5/P